T0309446

Hazardous Air Emissions from Incineration

Hazardous Air Emissions from Incineration

Calvin R. Brunner, P.E.

Chapman and Hall
New York • London

First published 1985
by Chapman and Hall
29 West 35th St. New York, NY 10001

Published in Great Britain by
Chapman and Hall Ltd
11 New Fetter Lane, London EC4P 4EE

Printed in the United States of America

Library of Congress Cataloging in Publication Data

Brunner, Calvin R.
Hazardous air emissions from incineration.

Bibliography: p.
Includes index.
1. Incineration—Environmental aspects—Handbooks, manuals, etc. 2. Air—Pollution—Environmental aspects—Handbooks, manuals, etc. 3. Hazardous wastes—Incineration—Environmental aspects—Handbooks, manuals, etc. I. Title.
TD796.B778 1985 628.5'32 84-27485
ISBN 0-412-00721-5

Dedicated to my parents,
David Brunner and Bella Brunner

Table of Contents

Foreword

This is a comprehensive handbook on the relationship of air pollution to incineration. Incineration is becoming the predominant method of dealing with many of our waste products and its most significant environmental impact is on the air.

This book includes information on emissions as well as on equipment design. Two chapters deal with the regulations governing incinerator emissions as well as the thermal destruction of hazardous wastes. Four chapters describe the nature of the emissions generated by the incineration process. These particulate, gaseous, and odor emissions, are hazardous as well as deleterious to public well-being and aesthetics. Also included is a complete and timely discussion of dioxin generation and discharges.

Three chapters describe the incineration equipment in general use today and methods of calculating gas flows and air discharges from these systems.

Five chapters discuss the types of gas cleaning equipment available with sizing information and expected efficiencies. The nature of the gas cleaning process is discussed in detail. Criteria for selection of the optimum system for a particular application is also included.

The dispersion of an atmospheric discharge to the surrounding areas and/or communities is a vital concern in assessing the nature of that discharge and its impact, or potential hazards. A chapter is devoted to a relative simple method of estimating atmospheric dispersion.

Noise discharge can be hazardous as well as a public nuisance. There is a chapter on the nature of noise emissions, noise control, and the impact of noise on individuals and communities.

All in all, this single text provides the reader with the information required to estimate incinerator emissions, determine its severity (or degree of hazard), and to choose that control equipment which provides the necessary remediation or control.

Calvin R. Brunner, P.E.
Reston, Va.
January 1985

List of Tables

Chapter 1

Chapter 2

Chapter 3

Chapter 4

Chapter 5

Chapter 6

Chapter 7

Chapter 9

Chapter 10

Chapter 11

Chapter 12

Chapter 13

Chapter 14

Chapter 15

Chapter 16

Chapter 17

Chapter 18

List of Figures

Chapter 2

Chapter 3

Chapter 4

Chapter 6

Chapter 7

Chapter 8

Chapter 10

Chapter 11

Chapter 12

Chapter 13

Chapter 14

Chapter 15

Chapter 17

Chapter 18

1

Introduction

Air is unique to our planet in the solar system. It is first of all a cover, providing protection from the short-wave radiation generated from extraterrestrial forces throughout our galaxy. Filtering out most of the cosmic and ultraviolet radiation bombarding the Earth, it is responsible for the development of the life that we have on our planet. It allows transmission of a limited amount of sunlight and solar heat, providing the relatively narrow range of temperatures that can support life as we know it.

Air is basically a mixture of two gases, nitrogen and oxygen, and the ratio of these two elements is relatively constant in all parts of the Earth. Air normally is found to contain traces of other elements, which will vary from place to place depending on topography, time of day, and season. Table 1–1 lists the relative values of the various components which make up a typical volume of air.

Air is, in a sense, as alive as the life which it supports. Plant life, in its diurnal cycle, will ingest the carbon dioxide within the air to generate

Table 1–1 Composition of Clean, Dry Air Near Sea Level

Component	% By Volume	ppm
Nitrogen	78.09	780,900.
Oxygen	20.94	209,400.
Argon	.93	9,300.
Carbon Dioxide	.0318	318.
Neon	.0018	18.
Helium	.00052	5.2
Methane	.00015	1.5
Krypton	.0001	1.0
Hydrogen	.00005	.5
Nitrous Oxide	.000025	.25
Carbon Monoxide	.00001	.1
Xenon	.000008	.08
Ozone	.000002	.02
Ammonia	.000001	.01
Nitrogen Dioxide	.0000001	.001
Sulfur Dioxide	.00000002	.0002

ppm = parts per million parts, by volume.

Source: L. Pryde, *Environmental Chemistry*, 1st ed. (Menlo Park, CA, Cummings, 1973).

oxygen in its process of producing carbohydrates, a process we know as photosynthesis. On the other hand, animal life breathes oxygen, producing the carbon dioxide needed by the plants.

Pollution is defined as the presence of matter or energy whose nature, location, or quantity produces undesirable environmental effects. While there may be controversy regarding classifying certain materials as pollutants within the air environment, general agreement exists over the undesirable effects of many gaseous and aerosol components found within the atmosphere. Table 1–2 lists the three major classes of pollutants found in the air. Inorganic gases are basically oxides, simple chemical compounds which are formed by the combination of an element, such as sulfur or nitrogen, with oxygen. Also included within this classification are gases containing halogens, nitrogen, or sulfur but which contain no carbon.

Carbon has a unique chemical property, covalence, common to only a few other elements, such as silicon, whereby it can develop a chemical

Table 1–2 Classification of Air Pollutants

Major Classes	Subclasses	Typical Members of the Subclasses
Inorganic Gases	Oxides of Nitrogen	Nitric Oxide, Nitrogen Dioxide
	Oxides of Sulfur	Sulfur Dioxide, Sulfur Trioxide
	Oxides of Carbon	Carbon Monoxide, Carbon Dioxide*
	Other Inorganics	Hydrogen Sulfide, Hydrogen Fluoride, Ammonia, Chlorine
Organic Gases	Hydrocarbons	Methane, Butane, Octane, Benzene, Acetylene, Ethylene, Butadiene
	Aldehydes and Ketones	Formaldehyde, Acetone
	Other Organics	Chlorinated Hydrocarbons, Organic Acids, Benzo-(a)-pyrene, Alcohols
Particulates	Solid Particulate	Fume, Dust, Smoke, Ash, Carbon, Lead
	Liquid	Mist, Spray, Oil, Grease, Acids

*Carbon Dioxide is considered a pollutant where it exists in greater concentrations than are found in the standard atmosphere.

Note: Gases, particularly organic gases, are normally considered to be compounds in gaseous state at standard temperature and pressure conditions. All other compounds which would be liquid at standard temperature and pressure are classified as "vapors," or "organic vapors."

Source: L. Pryde, *Environmental Chemistry*, 1st ed. (Menlo Park, CA: Cummings, 1973), 113.

bond with itself; that is, one carbon atom can form a stable bond with another carbon atom. This mechanism was once thought to be associated solely with organic processes and although it was found that covalent carbon compounds can be found without the presence of organisms, compounds containing covalent carbon are termed organic compounds. These compounds behave in a manner different from inorganic materials and they are, therefore, listed separately. (Note that not all carbon compounds are organic. For instance, carbon dioxide and carbon monoxide both contain carbon, but carbon is not present in covalent combination. These compounds are not classified as organic.)

Organic gases, those containing compounds of carbon where carbon is combined in covalent associations, are often characterized by odor and chemical reactivity. As will be discussed in Chapter 5, they are a major contributor to gross atmospheric disturbances such as smog, and some organics have been identified as carcinogens.

Particulates, or aerosols, encompass solid and liquid particles suspended in the air. They are identified with visible effects such as smoke

Table 1–3 Correlating Criteria Air Pollutants With Health

Pollutant	Effect
Ozone	Induces coughing, chest discomfort, and irritation of the nose, throat, and trachea: it aggravates asthma, emphysema, and chronic bronchitis: it causes chromosome breakdown in laboratory animals and hemolytic anema in humans: it acts on the human immunological system to reduce its resistance to bacterial infections, and on the neurological system to reduce the ability to concentrate.
Sulfur Dioxide	At high ambient levels it causes chemical bronchitis and may accentuate viral pneumonia. Small (0.5μ to 5μ) carbon particles which generally accompany sulfur dioxide pollution, can eventually produce emphysema and pulmonary fibrosis.
Carbon Monoxide	Produces reversible neurological effects such as headaches and dizziness. Chronic exposure to this odorless, colorless gas can lead to chronic oxygen insufficiency.
Nitrogen Dioxide	Among other things, it reduces the ability of the lung to cleanse itself of particulate matter. This gas destroys cilia and suppresses alveolar macrophage activity, the lung's final defense against foreign matter.
Hydrocarbons	These produce eye irritation. They have been indicted by some as possible causes of lung cancer.

and haze. There is some evidence to suggest that particles of a certain size range are absorbed within the lungs and will produce serious respiratory diseases.

As will be discussed in chapter 3, "National Ambient Air Quality Standards," six pollutants are currently regulated as "criteria" air pollutants by the federal government. Table 1–3 lists five of these pollutants with a discussion of the effects of each on the human body. Each of these pollutants beside ozone is a "primary" pollutant; that is, discharged from a stack. Ozone (O_3) is not a process emission but is produced by the action of sunlight on another pollutant, or by the interaction of one primary pollutant with another. Therefore, ozone is considered a "secondary" pollutant. The generation of ozone is discussed further in chapter 5.

The presence of any of the criteria pollutants in significant quantities is a cause for serious concern. Their effects range from eye irritation to lung cancer, as noted in Table 1–3.

There are many sources of air pollution in today's world, from natural as well as from nonnatural sources. It is estimated that over 260 tons of pollutants enter the atmosphere over the United States every year, as listed in Table 1–4. Transportation and power plants contribute the greatest amount of pollutant discharges through the burning of fossil fuels: coal, oil, and natural gas. The contribution of incineration is estimated at less than 5% of the total emissions discharge.

During the past decade efforts have been made by the federal government to reduce these figures. For instance, carbon monoxide and hydrocarbon emissions are generated from the incomplete combustion of gasoline and diesel fuel in vehicle cylinders. Catalytic convertors and exhaust gas recyclers have been mandated by statute. These mechanisms promote the effective burning of unburned carbon monoxide and hydrocarbon products, producing a much cleaner exhaust.

Table 1–4 Estimated Nationwide Emissions Millions of Tons Per Year

Source	Carbon Monoxide	Particulate	Sulfur Oxides	Hydrocarbons	Nitrogen Oxides	Total
Transportation	111.1	0.7	1.0	19.5	11.7	144.0
Power Plants	0.8	6.8	26.5	0.6	10.0	44.7
Industry	11.4	13.1	6.0	5.5	0.2	36.2
Incineration	7.2	1.4	0.1	2.0	0.4	11.1
Miscellaneous*	16.8	3.4	0.3	7.1	0.4	28.0
Total	147.3	25.4	33.9	34.7	22.7	264.0

*Primarily forest fires, agricultural burning, coal waste fires.

Source: *Environmental Quality*, The Third Annual Report of the Council on Environmental Quality (Washington, DC: Government Printing Office).

Likewise, the use of coal as a fuel for stationary power plants has been controlled by federal statute. The generation of sulfur dioxide derives from the burning of coal or residual oil with a high sulfur content. These laws limit the burning of high sulfur fuels in order to decrease the emission of sulfur oxides.

To examine the specific sources of nitrogen oxide (NO_x) emissions in the United States, refer to Table 1–5. Solid waste disposal, that is, incineration, produces less than 2% of the estimated nationwide emissions. Likewise, there are few states specifically regulating the emissions of nitrogen oxides from incineration processes.

Benzo(a)pyrene is a known carcinogen that is relatively easy to detect and that has been found to be almost universally produced from the combustion of organic material. It is therefore used as a measure of dangerous hydrocarbon emissions from a facility. As shown in Table 1–6, controlled burning discharges less than 34 tons a year into the atmosphere. In contrast, agricultural burning and forest fires generate over four times the quantity of benzo(a)pyrene than is produced by incineration. Coal waste fires, which are random and uncontrolled, produce over double the emissions of forest fires.

Table 1–5 Estimated Annual NO_x Emissions in the United States

Source	Estimated Emissions 1,000 Tons/Yr	Percent
Mobile		
Motor Vehicles	9096.	40.0
Aircraft	364.	1.6
Railroads	143.	0.6
Marine	165.	0.7
Nonhighway	1929.	8.5
Stationary		
Electric Utilities	4708.	20.7
Industrial Combustion	4531.	19.9
Commercial	221.	1.0
Residential	562.	2.5
Solid Waste Disposal	397.	1.8
Agricultural Burning	276.	1.2
Industrial Processes	199.	0.9
Miscellaneous	143.	0.6
Total	22,734.	100.0

Source: "Medical and Biological Effects of Environmental Pollutants," *Nitrogen Oxides*. (Washington, DC: National Academy of Sciences, 1977): 22.

Table 1–6 Summary of Benzo(a)Pyrene Emissions by Stationary Source in the U.S.A.

Source		Benzo(a)pyrene Emissions, Tons/Yr
Heat and Power Generation		500.
Coke Production		200.
Waste Burning:		
Enclosed Incinerators		
Municipal	<1.	
Commercial and Industrial	23.	
Institutional	2.	
Apartment	8.	
Open Burning		
Municipal	4.	
Commercial and Industrial	10.	
Domestic	10.	
Forest and Agricultural	140.	
Vehicle Disposal	50.	
Coal Refuse Fires	340.	
Total Waste Burning		588
Total Annual Release to the Atmosphere		1288

Source: *Biological Effects of Atmospheric Pollutants, Particulate Polycyclic Organic Matter.* (Washington, DC: National Academy of Sciences, 1982).

While incineration produces less significant discharges than many other combustion processes, these discharges are perceived by the public as severe nuisance emissions. The burning of refuse until the past decade has been associated with smoke and odor. In cities, apartment houses normally had their own incinerators and the residents were acutely aware of the presence of these incinerators when they were fired up. Legislation has effectively banned most of these incinerators through the imposition of air pollution control standards over the past decade, but the negative public attitude toward incineration still stands.

Refuse draws the most attention of all of the wastes generated in this country because everybody generates it. Looking at all of the solid waste produced, refuse is a small percentage of the total. Almost 80% of the solid waste generated in the United States, as shown in Table 1–7, comes from animal farming (manure, food processing wastes) and mining (overburden, slag from ore processing). Crop wastes, such as bagasse and straw, total three times the quantity of refuse, or municipal solid waste.

Table 1–7 Solid Wastes Originated From Certain Resource Categories

Source	Million Tons/Yr	Percent
Municipal	230	5.2
Industrial	140	3.1
Mineral	1700	38.2
Animal Wastes	1740	39.1
Crop Wastes	640	14.4
Total	4450	100.0

Source: U.S. Environmental Protection Agency, *Report to U.S. Congress—Resource, Recovery and Source Reduction, U.S./EPA Report SW-118* (Washington, DC: Government Printing Office, February 1978).

To place these numbers in perspective, note that municipal waste must be collected and is normally disposed of in a place remote from where it is generated. Animal waste is often used as fertilizer, mining wastes are normally left where they are generated, and crop wastes are either plowed under or are disposed of at the processing plant; for example, bagasse is usually fired in a specialty boiler at a sugar mill as fuel for the generation of steam. Industry constantly examines means of reusing their waste products. The disposal of refuse, therefore, receives the most public attention, with the exception of sporadic concern for the disposal of hazardous or nuclear waste materials.

References and Bibliography

1. American Industrial Hygiene Association, *Community air quality guides*, Westmont, NJ: AIHA, undated.
2. Air pollution control. *Power Magazine* Special Report (June 1976).
3. Balakrishnan, N., C. Cheng, and M. Patel. 1979. Emerging technologies for air pollution control. *Pollution Engineering*, (November).
4. Jacob, W. 1980. Reconciling clean air with energy demand. *Consulting Engineer* (April).
5. National Association of Manufacturers. 1975. *Air Quality Control*, New York: NAM, 1975.

2

Statutory Requirements

The Clean Air Act (CAA) passed by the federal government in 1970 was landmark legislation. It reflected the public's concern for the quality of the air environment and it led the way for state and local action in cleaning up the air. There have been significant changes to the CAA since its initial passage, and additional legislation has been passed by federal and state governments relative to incinerator emissions. In this chapter the more important of these regulations will be presented and described.

Federal Regulations

The federal government has established an intensive program for the regulation of hazardous waste incineration. Nonhazardous waste incineration is regulated by the federal government only if the waste is generated from municipal sources. Nonhazardous waste generated and incinerated by industry is not at present regulated by the federal government except for the provisions of the National Ambient Air Quality Standards (NAAQS) and Prevention of Significant Deterioration (PSD) regulations, both of which will be discussed subsequently.

Municipal Waste Incineration

Emissions from municipal solid waste incinerators are covered by the New Source Performance Standards, Title 40, Part 60, Subpart E. Municipal solid waste is defined as refuse, more than 50% of which is muncipal-type waste consisting of a mixture of paper, wood, yard wastes, plastics, leather, rubber, and other combustibles, and noncombustibles such as glass and rock. This standard requires that air emissions not exceed a rate of 0.08 grains of particulate per dry standard cubic foot of exhaust gas, corrected to 12% carbon dioxide.

The second federal regulation established for control of emissions from municipal incinerators is subpart O of the above standard. This statute establishes the emissions from incinerators burning sewage sludge generated from sewage treatment facilities. It limits particulate emissions from the facility to 1.3 pounds per dry ton of sludge charged. It also limits visible emissions to 20% opacity.

Solid waste incinerator standards apply to incinerators constructed after August 17, 1971. The sewage sludge incinerator standards apply to incinerators constructed after June 11, 1973.

Definition of Hazardous Waste

The United States Environmental Protection Agency, through regulation 40 CFR Part 261, dated May 19, 1980, of the Resource Conservation and Recovery Act (RCRA), and subsequent revisions, has established a set of criteria in an attempt to define a waste as a hazardous waste, as follows:

- Ignitability: A liquid having a flash point less than 140°F: a substance other than a liquid which can cause fire through friction, absorbtion of moisture, or spontaneous chemical changes under standard temperature and pressure, certain flammable solids and explosive gases.
- Corrosivity: An aqueous waste with pH equal to or less than 2.0 or a pH equal to or greater than 12.5: a liquid that corrodes carbon steel at a rate greater than 0.250 inches per year.
- Reactivity: A substance which is normally unstable and undergoes violent physical and/or chemical change without detonating; a substance that reacts violently with water; a substance which can generate harmful gases, vapors, or fumes when mixed with water; a substance which is readily capable of detonation at standard temperature and pressure, etc.
- Extraction Procedure Toxicity (EP Toxicity): If the extract (or leachate) from a representative sample of this waste contains contamination in excess of allowable levels, it is considered a hazardous waste. The allowable level is defined as up to 10 times that allowed in drinking water for the contaminant in question.
- Acute Hazardous Waste: A substance which has been found to be fatal to humans in low doses, or in the absence of data on human toxicity, has been found to be fatal in corresponding human concentrations in laboratory animals.
- Toxicity: Wastes that have been found, through laboratory studies, to have a carcinogenic (producing or tending to produce cancer), mutagenic (capable of inducing mutations in future offspring), or tetratogenic (producing abnormal growth in fetuses) effect on human or other life forms.

A set of lists have been included in this regulation identifying specific hazardous wastes based on the above definitions, such as industries or processes where the waste generated must always be considered hazardous. One of these lists, Appendix 8 of this regulation, "Hazardous

Constituents," identifies hundreds of compounds which are known to be hazardous based on the above criteria.

Principal Organic Hazardous Constituent

The hazardous waste incineration regulations require that the principal organic hazardous constituent (POHC) of a waste be defined. Where a waste contains one or more of the hazardous constituents listed in Appendix 8 of the regulations, reproduced as Table 2–1, the regulatory agency (EPA) must determine which of these components is the principal component. This decision is based upon the percentage of each component present, the degree of hazard of the component, and its persistence. If none of these components are present the waste is not subject to the hazardous waste incinerator regulations. (Appendix 8 will be reviewed and updated on a regular basis. It contains the more common hazardous compounds found in industry today and it will eventually list a relatively large percentage of the thousands of hazardous components generated today and anticipated for the future).

Incineration of Hazardous Wastes

The POHC must be identified. The incinerator must be designed and operated to achieve a destruction and removal efficiency (DRE) of at least 99.99% (four nines destruction). The DRE is defined as follows, with W_{in} the POHC mass rate into the incinerator and W_{out} the POHC mass rate leaving in the incinerator exhaust gas discharge to atmosphere:

$$DRE = \frac{W_{in} - W_{out}}{W_{in}} \cdot 100$$

If the hazardous waste generates in excess of 4 lb/hr of hydrogen chloride in the exhaust gas stream of an incinerator, 99% of the HCl produced must be removed prior to discharge into the atmosphere. (This regulation in essence mandates the use of scrubbers for such applications.)

Particulate emissions into the atmosphere must not exceed 0.08 grains per dry standard cubic foot when corrected to 50% excess air.

Additional requirements contained within this regulation include a test burn to establish incinerator parameters for particular waste, and an extensive monitoring requirement.

Polychlorinated Biphenyl Waste

Wastes containing polychlorinated biphenyls (PCBs) in a concentration greater than 50 parts per million by weight are presently covered by the Toxic Substances Control Act (TSCA) 44 CFR 106 paragraph 761.41, not

Table 2.1 Hazardous Constituents

Acetaldehyde
(Acetato)phenylmercury
Acetonitrile
3-(alpha-Acetonylbenzyl)-4-hydroxycoumarin and salts
2-Acetylaminofluorene
Acetyl chloride
1-Acetyl-2-thioures
Acrolein
Acrylamide
Acrylonitrile
Aflatoxins
Aldrin
Allyl alcohol
Aluminum phosphide
4-Aminobiphenyl
6-Amino-1.1a,2.8.8a.8b-hexahydro-8-(hydroxymethyl)-8a-methoxy-5-methylcarbamate azirino(2′.3′:3.4) pyrrolo(1.2-a)indole-4.7-dione (ester) (Mitomycin C)
5-(Aminomethyl)-3-isoxazolal
4-Aminopyridine
Amitrole
Antimony and compounds, N.O.S.[1]
Aramite
Arsenic and compounds, N.O.S.
Arsenic acid
Arsenic pentoxide
Arsenic trioxide
Auramine
Azaserine
Barium and compounds, N.O.S.
Barium cyanide
Benz(c)acridine
Benz(a)anthracene
Benzene
Benzenearsonic acid
Benzenethiol
Benzidine
Benzo(a)anthracene
Benzo(b)fluoranthene
Benzo(j)fluoranthene
Benzo(a)pyrene
Benzotrichloride
Benzyl chloride
Beryllium and compounds, N.O.S.
Bis(2-chloroethyoxy)methane
Bis(2-chloroethyl)ether
N,N-Bis(2-chloroethyl)-2-naphthylamine
Bis(2-chloroisopropyl) ether
Bis(chloromethyl) ether
Bis(2-ethylhexyl) phthalate
Bromoacetone
Bromomethane
4-Bromophenyl phenyl ether
Brucine
2-Butanone peroxide
Butyl benzyl phthalate
2-sec-Butyl-4.6-dinitrophenol (DNBP)
Cadmium and compounds, N.O.S.
Calcium chromate
Calcium cyanide
Carbon disulfide
Chlorambucil
Chlordane (alpha and gamma isomers)
Chorlinated benzenes, N.O.S.
Chlorinated ethane, N.O.S.
Chlorinated naphthalene, N.O.S.
Chlorinated phenol, N.O.S.
Chloroacetaldehyde
Chloroalkyl ethers
p-Chloroaniline
Chlorobenzene
Chlorobenzilate
1-(p-Chlorobenzoyl)-5-methoxy-2-methylindole-3-acetic acid
p-Chloro-m-cresol
1-Chloro-2,3-epoxybutane
2-Chloroethyl vinyl ether
Chloroform
Chloromethane
Chloromethyl methyl ether
2-Chloronaphthalene
2-Chlorophenol
1-(o-Chlorophenyl)thioures
3-Chloropropionitrile
alpha-Chlorotoluene
Chlorotoluene, N.O.S.
Chromium and compounds, N.O.S.
Chrysene
Citrus red No. 2
Copper cyanide
Creosote
Crotonaldehyde

Table 2–1 Hazardous Constituents (*Continued*)

Cyanides (soluble salts and complexes), N.O.S.
Cyanogen
Cyanogen bromide
Cyanogen chloride
Cycasin
2-Cyclohexyl-4.6-dinitrophenol
Cyclophosphamide
Daunomycin
DDD
DDE
DDT
Diallate
Dibenz(a.h)acridine
Dibenz(a,j)acridine
Dibenz(a,h)anthracene(Dibenzo[a.h] anthracene)
7H-Dibenzo(c.g)carbazole
Dibenzo(a,e)pyrene
Dibenzo(a,i)pyrene
1.2-Dibromo-3-chloropropane
1.2-Dibromoethane
Dibromomethane
Di-n-butyl phthalate
Dichlorobenzene, N.O.S.
3.3′-Dichlorobenzidine
1.1-Dichloroethane
1.2-Dichloroethane
trans-1.2-Dichloroethane
Dichloroethylene, N.O.S.
1.1-Dichloroethylene
Dichloromethane
2.4-Dichlorophenol
2.6-Dichlorophenol
2.4-Dichlorophenoxyacetic acid (2.4-D)
Dichloropropane
Dichlorophenylarsine
1.2-Dichloropropane
Dichloropropanol, N.O.S.
Dichloropropene, N.O.S.
1.3-Dichloropropene
Dieldrin
Diepoxybutane
Diethylarsine
0,0-Diethyl-S-(2-ethylthio)ethyl ester of phosphorothioic acid
1.2-Diethylhydrazine
0,0-Diethyl-S-methylester phosphorodithioic acid
0,0-Diethylphosphoric acid, 0-p-nitrophenyl ester
Diethyl phthalate
0,0-Diethyl-0-(2-pyrazinyl)phosphorothioate
Diethylstilbestrol
Dihydrosafrole
3.4-Dihydroxy-alpha-(methylamino)-methyl benzyl alcohol
Di-isopropylfluorophosphate (DFP)
Dimethoate
3.3′-Dimethyoxybenzidine
p-Dimethylaminoazobenzene
7.12-Dimethylbenz(a)anthracene
3.3′-Dimethylbenzidine
Dimethylcarbamoyl chloride
1.1-Dimethylhydrazine
1.2-Dimethylhydrazine
3.3-Dimethyl-1-(methylthio)-2-butanone-0-((methylamino) carbonyl)oxime
Dimethylnitrosoamine
alpha,alpha-Dimethylphenethylamine
2.4-Dimethylphenol
Dimethyl phthalate
Dimethyl sulfate
Dinitrobenzene, N.O.S.
4.6-Dinitro-o-cresol and salts
2.4-Dinitrophenol
2.4-Dinitrotoluene
2.6-Dinitrotoluene Di-n-octyl phthalate
1.4-Dioxane
1.2-Diphenylhydrazine
Di-n-propylnitrosamine
Disulfoton
2.4-Dithiobiuret
Endosulfan
Endrin and metabolites
Epichlorohydrin
Ethyl cyanide
Ethlene diamine
Ethylenebisdithiocarbamate (EBDC)
Ethyleneimine
Ethylene oxide
Ethylenethiourea
Ethyl methanesulfonate
Fluoranthene
Fluorine
2-Fluoroacetamide

Table 2–1 Hazardous Constituents (*Continued*)

Fluoroacetic acid, sodium salt
Formaldehyde
Glycidylaldehyde
Halomethane, N.O.S.
Heptachlor
Heptachlor epoxide (alpha, beta, and gamma isomers)
Hexachlorobenzene
Hexachlorobutadene
Hexachlorocyclohexane (all isomers)
Hexachlorocyclopentadiene
Hexachloroethane
1.2.3.4.10.10-Hexachlora-1.4.4a,5.8.8a-hexahydro-1.4:5.8-endo.endo-dimethanonaphthalene
Hexachlorophene
Hexachloropropene
Hexaethyl tetraphosphate
Hydrazine
Hydrocyanic acid
Hydrogen sulfide
Indeno(1.2.3-c.d)pyrene
Iodomethane
Isocynanic acid, methyl ester
Isosafrole
Kepone
Lasiocarpine
Lead and compounds, N.O.S.
Lead acetate
Lead phosphate
Lead subacetate
Maleic anhydride
Malononitrile
Melphalan
Mercury and compounds, N.O.S.
Methapyrilene
Methomyl
2-Methylaziridine
3-Methylcholanthrene
4.4′-Methylene-bis-(2-chloroaniline)
Methyl ethyl ketone (MEK)
Methyl hydrazine
2-Methyllactonitrile
Methyl methacrylate
Methyl methanesulfonate
2-Methyl-2-methylthio)propionaldehyde-o-(methylcarbonyl) oxime
N-Methyl-N′-nitro-N-nitrosoguanidine
Methyl parathion
Methylthiouracil
Mustard gas
Naphthalene
1.4-Naphthoquinone
1-Naphthylamine
2 Naphthylamine
1-Naphthyl-2-thioures
Nickel and compounds,N.O.S.
Nickel carbonyl
Nickel cyanide
Nicotine and salts
Nitric oxide
p-Nitroaniline
Nitrobenzene
Nitrogen dioxide
Nitrogen mustard and hydrochloride salt
Nitrogen mustard N-oxide and hydrochloride salt
Nitrogen peroxide
Nitrogen tetroxide
Nitroglycerine
4-Nitrophenol
4-Nitroquinoline-1-oxide
Nitrosamine, N.O.S.
N-Nitrosodi-N-butylamine
N-Nitrosodiethanolamine
N-nitrosodiethylamine
N-Nitrosodimethylamine
N-Nitrosodiphenylamine
N-Nitrosodi-N-propylamine
N-Nitroso-N-ethylurea
N-Nitrosomethylethylamine
N-Nitroso-N-methylurea
N-Nitroso-N-methyurethane
N-Nitrosomethylvinylamine
N-Nitrosomorpholine
N-Nitrosonornicotine
N-Nitrosopiperidine
N-Nitrosopyrrolidine
N-Nitrosoarcosine
5-Nitro-o-toluidine
Octamethylpyrophosphoramide
Oleyl alcohol condensed with 2 moles ethylene oxide
Osmium tetroxide
7-Oxabicyclo(2.2.1)heptane-2.3-dicarboxylic acid

Table 2–1 Hazardous Constituents (*Continued*)

Parathion
Pentachlorobenzene
Pentachloroethane
Pentachloronitrobenzene (PCNB)
Pentacholorophenol
Phenacetin
Phenol
Phenyl dichloroarsine
Phenylmercury acetate
N-Phenylthiourea
Phosgene
Phosphine
Phosphorothioic acid, 0.0-dimethyl ester, O-ester with N.N-dimethyl benzene sulfonamide
Phthalic acid esters, N.O.S.
Phthalic anhydride
Polychlorinated biphenyl, N.O.S.
Potassium cyanide
Potassium silver cyanide
Pronamide
1.2-Propanediol
1.3-Propane sultone
Propionitrile
Propylthiouracil
2-Propyn-1-ol
Pryidine
Reserpine
Saccharin
Sefrole
Selenious acid
Selenium and compounds, N.O.S.
Selenium sulfide
Selenourea
Silver and compounds, N.O.S.
Silver cyanide
Sodium cyanide
Streptozotocin
Strontium sulfide
Strychnine and salts
1.2.4.5-Tetrachlorobenzene
2.3.7.8-Tetrachlorodibenzo-p-dioxin (TCDD)
Tetrachloroethane, N.O.S.
1.1.1.2-Tetrachloroethane
1.1.2.2-Tetrachloroethane
Tetrachloroethane (Tetrachloroethylene)
Tetrachloromethane
2.3.4.6-Tetrachlorophenol
Tetraethyldithiopyrophosphate
Tetraethyl lead
Tetraethylpyrophosphate
Thallium and compounds, N.O.S.
Thallic oxide
Thallium (I) acetate
Thallium (I) carbonate
Thallium (I) chloride
Thallium (I) nitrate
Thallium selenite
Thallium (I) sulfate
Thioacetamide
Thiosemicarbazide
Thiourea
Thiuram
Toluene
Toluene diamine
o-Toluidine hydrochloride
Tolylene diisocyanate
Toxaphene
Tribromomethane
1.2.4-Trichlorobenzene
1.1.1-Trichloroethane
Trichloroethene (Trichloroethylene)
Trichloromethanethiol
2.4.5-Trichlorophenol
2.4.6-Trichlorophenol
2.4.5-Trichlorophenoxyacetic acid (2.4.5-T)
2.4.5-Trichlorophenoxypropionic acid (2.4.5-TP) (Silvex)
Trichloropropane, N.O.S.
1.2.3-Trichloropropane
0,0,0-Triethyl phosphorothioate
Trinitrobenzene
Tris(1-azridinyl)phosphine sulfide
Tris(2.3-dibromopropyl) phosphate
Trypan blue
Uracil mustard
Urethane
Vanadic acid, ammonium salt
Vanadium pentoxide (dust)
Vinyl chloride
Vinylidene chloride
Zinc cyanide
Zinc phosphide

[1]The abbreviation N.O.S. signifies those members of the general class "not otherwise specified" by name in this listing.

RCRA. The persistence and potential danger of PCB materials in the environment was one of the first issues to attract public attention in the field of hazardous waste disposal. Regulations governing the disposal of PCBs were promulgated in response to this concern, years before passage of RCRA. It is expected that PCB waste disposal will eventually be included under the RCRA regulations. At this time, however, disposal of PCBs by incineration requires the following parameters of operation and design:

- Combustion of the material at 2192°F with a 2-second retention time and 3% excess oxygen in the exhaust gas, or at 2912°F, with a 1½-second retention time and 2% excess air in the exhaust.
- Combustion efficiency (CE) of 99.9%. With C_{co_2} and C_{co} the concentrations of carbon dioxide and carbon monoxide in the exhaust gas, respectively, the CE is calculated as follows:

$$CE = \frac{C_{co_2}}{C_{co_2} + C_{co}} \cdot 100$$

- Water scrubbers or equivalent gas cleaning equipment shall be installed to control hydrogen chloride emissions in the exhaust gas.
- If the PCBs to be incinerated are nonliquid, in addition to the above requirements the mass air emmissions shall not be greater than 1 pound PCB per million pounds charged into the furnace (99.9999% or six nines destruction.)

Additional PCB incineration regulations address process monitoring and operating safety.

State Particulate Emission Standards

Emission standards vary from state to state in both qualitative and quantitative values. Standards are expressed in grains per cubic foot, in pounds per unit weight charged, or in other units. Table 2–2 relates these emission factors to one another. Note that the conversion of one factor to another is dependent upon the nature of the waste being burned. For instance, if the waste were rich in carbon, adjustment to 12% carbon dioxide would be different than if the waste were high in cellulose materials. This table is based upon a typical refuse composition and is not applicable to substances whose chemical composition is greatly different from that of refuse.

Table 2–3 is a listing of the particulate emissions regulations for each of the 50 states. The last column of this chart is an equivalent discharge in grains per cubic foot, in accordance with the conversion factors listed in Table 2–2, which provides a common basis of comparison between one state and another. These standards apply to all incineration facilities;

Table 2–2 Conversion Factors

	Lb/Ton Refuse as Received	Lb/1000lb Flue Gas at 50% Excess Air	Lb/1000lb Flue Gas at 12% CO_2	GR/DSCF at 50% Excess Air	GR/DSCF at 12% CO_2	G/NCM at NTP 7% CO_2
Lb/Ton Refuse as Received	1	0.089	0.010	0.047	0.053	0.067
Lb/1000 Lb Flue Gas at 50% Excess Air	11.27	1	1.12	0.52	0.585	0.74
Lb/1000 Lb Flue Gas at 12% CO_2	10.0	0.89	1	0.46	0.52	0.66
GR/DSCF at 50% Excess Air	21.31	1.93	2.16	1.12	1	1.42
GR/DSCF at 12% CO_2	18.85	1.71	1.92	0.89	1	1.26
G/NCM at NTP 7% CO_2	15.0	1.36	1.53	0.704	0.79	1

Note: This table is based upon a typical refuse composition: 8.2% metals, 35.6% paper, 11% plastics, 1.5% leather and rubber, 1.9% textiles, 2.5% wood, 23.7% food waste (garbage), 15.5% yard waste, 8.3% glass and 1.7% miscellaneous waste, 4462 Btu/lb as charged.

Example: 1 lb/ton refuse is equivalent to 0.053 grain/dscf at 12% CO_2: 1 G/NCM at NTP, 7% CO_2, is equivalent to 1.36 lb/1000 lb flue gas at 50% excess air.

Source: W. Niessen and A. Sarofin, *Incinerator Air Pollution: Facts and Speculation*, In: *Proceedings* of the ASME 1970 National Incinerator Conference (New York: ASME, 1970).

Table 2–3 Particulate Emission Limitations for New and Existing Incinerators

	Regulation					Equivalent
State	Value	Units	Corrected to	Process Conditions	Validity	Common Regulation (gr/dscf @ 12% CO_2)
1 Alabama	0.1	lbs/100 lbs charged		>50 TPD		0.12
	0.2	lbs/100 lbs charged		≤50 TPD		0.24
2 Alaska	0.3	gr/dscf	12% CO_2	≤200 lbs/hr		0.3
	0.2	gr/dscf	12% CO_2	200–1000 lbs/hr		0.2
	0.1	gr/dscf	12% CO_2	>1000 lbs/hr		0.1
3 Arizona	0.1	gr/dscf	12% CO_2			0.1
4 Arkansas	0.2	gr/dscf	12% CO_2	≥200 lbs/hr		0.2
	0.3	gr/dscf	12% CO_2	<200 lbs/hr	0.3	
5 California	0.3	gr/dscf	12% CO_2	typical of the 43 APCD's		0.3
6 Colorado	0.1	gr/dscf	12% CO_2		designated control areas	0.1
	0.15	gr/dscf	12% CO_2		other areas	0.15
7 Connecticut	0.08	gr/dscf	12% CO_2		built after 6/1/72	0.08
	0.4	lbs/1000 lbs	50% excess air		built before 6/1/72	0.26
8 Delaware	0.2	lbs/hr		100 lbs/hr		0.24
	1.0	lbs/hr		500 lbs/hr		0.24
	2.0	lbs/hr		1000 lbs/hr		0.24
	5.0	lbs/hr		3000 lbs/hr		0.2
9 Florida	0.08	gr/dscf	50% excess air	≥50 TPD	built after 2/11/72	0.1
	0.1	gr/dscf	50% excess air	≥50 TPD	built before 2/11/72	0.12
10 Georgia	0.1	gr/dscf	12% CO_2	≤50 TPD—type 0, 1, 2 waste	new (built after 1/1/72)	0.1
	0.2	gr/dscf	12% CO_2	≤50 TPD—type 3, 4, 5, 6 waste	new (built after 1/1/72)	0.2
	0.2	gr/dscf	12% CO_2	type 0,1,2 waste	existing before 1/1/72	0.2
	0.3	gr/dscf	12% CO_2	type 3, 4, 5, 6 waste	existing before 1/1/72	0.3
	0.08	gr/dscf	12% CO_2	≥50 TPD	new (built after 1/1/72)	0.08

Table 2–3 Particulate Emission Limitations for New and Existing Incinerators (*Continued*)

State	Regulation					Equivalent
	Value	Units	Corrected to	Process Conditions	Validity	Common Regulation (gr/dscf @ 12% CO_2)
11 Hawaii	0.2	lbs/100 lbs charged				0.24
12 Idaho	0.2	lbs/100 lbs charged				0.24
13 Illinois	0.08	gr/dscf	12% CO_2	2000–60,000 lbs/hr		0.08
	0.2	gr/dscf	12% CO_2	≤2000 lbs/hr	built before 4/15/72	0.2
	0.1	gr/dscf	12% CO_2	≤2000 lbs/hr	built after 4/15/72	0.1
14 Indiana	0.3	lbs/1000 lbs gas	50% excess air	≥200 lbs/hr		0.19
	0.5	lbs/1000 lbs gas	50% excess air	<200 lbs/hr		0.32
15 Iowa	0.2	gr/dscf	12% CO_2	≥1000 lbs/hr	0.2	
	0.35	gr/dscf	12% CO_2	<1000 lbs/hr		0.35
16 Kansas	0.3	gr/dscf	12% CO_2	<200 lbs/hr		0.3
	0.2	gr/dscf	12% CO_2	200–20,000 lbs/hr		0.2
	0.1	gr/dscf	12% CO_2	>20,000 lbs/hr		0.1
17 Kentucky	0.2	gr/dscf	12% CO_2	≤50 TPD		0.2
	0.08	gr/dscf	12% CO_2	>50 TPD	0.08	
18 Louisiana	0.2	gr/dscf	12% CO_2			0.2
19 Maine	0.2	gr/dscf	12% CO_2			0.2
20 Maryland	0.1	gr/dscf	12% CO_2	<2000 lbs/hr	built after 1/17/72	0.1
	0.03	gr/dscf	12% CO_2	>2000 lbs/hr	built after 1/17/72	0.03
	0.3	gr/dscf	12% CO_2	<200 lbs/hr	built before 1/17/72	0.3
	0.2	gr/dscf	12% CO_2	>200 lbs/hr	built before 1/17/72	0.2
21 Massachusetts	0.1	gr/dscf	12% CO_2		existing	0.1
	0.05	gr/dscf	12% CO_2		new	0.05
22 Michigan	0.65	lbs/1000 lbs gas	50% excess air	0–100 lbs/hr		0.42
	0.3	lbs/100 lbs gas	50% excess air	>100 lbs/hr		0.19
23 Minnesota	0.3	gr/dscf	12% CO_2	<200 lbs/hr	existing before 8/17/71	0.3
	0.2	gr/dscf	12% CO_2	200–2000 lbs/hr	existing before 8/17/71	0.2
	0.1	gr/dscf	12% CO_2	>2000 lbs/hr	existing before 8/17/71	0.1

	0.2	gr/dscf	12% CO_2	<200 lbs/hr	new (built after8/17/71)	0.2
	0.15	gr/dscf	12% CO_2	200–2000 lbs/hr	new (built after 8/17/71)	0.15
	0.1	gr/dscf	12% CO_2	>2000 lbs/hr	new (built after 8/17/71)	0.1
24 Mississippi	0.2	gr/dscf	12% CO_2	Design capacity		0.2
	0.1	gr/dscf	12% CO_2	New sources near residential areas		0.1
25 Missouri	0.2	gr/dscf	12% CO_2	⩾200 lbs/hr		0.2
	0.3	gr/dscf	12% CO_2	<200 lbs/hr		0.3
26 Montana	0.2	gr/dscf	12% CO_2	>200 lbs/hr	existing before 9/5/75	0.2
	0.3	gr/dscf	12% CO_2	⩽200 lbs/hr	existing before 9/5/75	0.3
	0.1	gr/dscf	12% CO_2		all others	0.1
27 Nebraska	0.2	gr/dscf	12% CO_2	<2000 lbs/hr		0.2
	0.1	gr/dscf	12% CO_2	⩾2000 lbs/hr		0.1
28 Nevada	3.0	lbs/ton charged		<2000 lbs/hr		0.18
	variable	$E = 40.7 \times 10^{-5}$ C	C,E = lbs/hr	>2000 lbs/hr		0.05
29 New Hampshire	0.3	gr/dscf	12% CO_2	⩽200 lbs/hr		0.3
	0.2	gr/dscf	12% CO_2	>200 lbs/hr		0.2
	0.08	gr/dscf	12% CO_2	>50 TPD	built after 4/20/74	0.08
30 New Jersey	0.2	gr/dscf	12% CO_2	<2000 lbs/hr	type 0, 1, 2, 3 waste only	0.2
	0.2	gr/dscf	12% CO_2	all others		0.1
31 New Mexico	only opacity	regulations		⩽50 TPD		—
	0.08	gr/dscf	12% CO_2	>50 TPD	new (built after 8/17/71)	0.08
32 New York	0.5	lbs/100 lbs charged		>2000 lbs/hr	built between 4/1/62 and 1/1/70	0.6
	0.5	lbs/100 lbs charged		⩽2000 lbs/hr	built between 4/1/62 and 1/1/68	0.6
	variable (e.g., 0.3)	lbs/hr		⩽100 lbs/hr	built after 1/1/68	0.36
	variable (e.g., 3.0)	lbs/hr		@1000 lbs/hr	built after 1/1/68	0.36
	variable (e.g., 7.5)	lbs/hr		@3000 lbs/hr	built after 1/1/70	0.3

Table 2–3 Particulate Emission Limitations for New and Existing Incinerators (*Continued*)

State	Regulation					Equivalent
State	Value	Units	Corrected to	Process Conditions	Validity	Common Regulation (gr/dscf @ 12% CO_2)
33 North Carolina	0.2	lbs/hr		0–100 lbs/hr		0.24
	0.4	lbs/hr		@200 lbs/hr		0.24
	1.0	lbs/hr		@500 lbs/hr		0.24
2.0	lbs/hr		@1000 lbs/hr		0.24	
	4.0	lbs/hr		≥2000 lbs/hr		0.24
34 North Dakota	variable	lbs/hr		@100 lbs/hr		0.4
			@1000 lbs/hr			0.31
			@3000 –/hr			0.24
35 Ohio	0.1	lbs/100 lbs charged		≥100 lbs/hr		0.12
	0.2	lbs/100 lbs charged		<100 lbs/hr		0.24
36 Oklahoma	variable	lbs/hr		@100 lbs/hr		0.48
				@1000 lbs/hr		0.31
				@3000 lbs/hr		0.21
37 Oregon	0.3	gr/dscf		≥100 lbs/hr		0.3
	0.2	gr/dscf		>200 lbs/hr	built before 6/1/70	0.2
	0.1	gr/dscf		>200 lbs/hr	built after 6/1/70	0.1
38 Pennsylvania	0.1	gr/dscf	12% CO_2			0.1
39 Rhode Island	0.16	gr/dscf	12% CO_2	<2000 lbs/hr		0.16
	0.08	gr/dscf	12% CO_3	≥2000 lbs/hr		0.08
40 South Carolina	0.5	lbs/10^6 Btu		@10 mm Btu/hr		0.27
41 South Dakota	0.2	lbs/100 lbs charged				
42 Tennesee	0.2	% of charge	≤2000 lbs/hr		0.24	
	0.1	% of charge	≥2000 lbs/hr		0.12	
43 Texas	variable	lbs/hr		@1000 lbs/hr		0.41
				@3000 lbs/hr		0.27
44 Utah	0.08	gr/dscf	12% CO_2	>50 TPD		0.08
45 Vermont	0.1	lbs/100 lbs charged				0.12

46 Virginia	0.14	gr/dscf	12% CO_2			0.14
47 Washington	0.1	gr/dscf	7% O_2			0.11
48 West Virginia	8.25	lbs/ton		≤200 lbs/hr		0.5
	5.43	lbs/ton		>200 lbs/hr		0.33
49 Wisconsin	0.2	lbs/1000 lbs exhaust gas	12% CO_2	500–4000 lbs/hr	built after 4/1/72	0.11
	0.3	lbs/1000 lbs exhaust gas	12% CO_2	≤500 lbs/hr	built after 4/1/72	0.17
	0.5	lbs/1000 lbs exhaust gas	12% CO_2	>500 lbs/hr	built before 4/1/72	0.28
	0.6	lbs/1000 lbs exhaust gas	12% CO_2	≤500 lbs/hr	built before 4/1/72	0.34
	0.15	lbs/1000 lbs exhaust gas	12% CO_2	≥4000 lbs/hr	built after 4/1/72	0.08
50 Wyoming	0.2	lbs/100 lbs charged				0.24

for example, municipal refuse, municipal sludge, industrial, institutional, and agricultural incinerators.

State Opacity Standards

Opacity is a visual measure of incinerator atmospheric discharge. In general, the standards governing particulate emissions have an effect on stack opacity. With opacity a visible quality of a discharge, however, it was incumbent upon public officials to establish a standard that can be easily identified and verified by lay people. Figure 2–1 is a copy of a typical opacity chart. The chart is normally printed on a transparent surface and the degree of opacity of the exhaust is matched to the chart reading to obtain a qualitative measure of degree of opacity.

Table 2–4 lists opacity requirements for each state for new and existing commercial and industrial incinerators. The percent opacity is that opacity as noted in Figure 2–1.

References and Bibliography

1. *NESHAPS, National emission standards for hazardous air pollutants.* 40 CFR 61. October 1981.
2. North Atlantic Treaty Organization. November 1976. *Air Quality Criteria for Particulate Matter.* Washington, DC: Government Printing Office, PB240/570.
3. Union Carbide Corporation. November 1978. *The Clean Air Act Amendments of 1977.*
4. Deland, M. A national contingency plan for hazardous waste. *Environmental Science and Technology* (May 1982).
5. Summary of Current RCRA Incinerator Regulations, *Hazardous Waste Report* (November 15, 1982).
6. Brunner, C. Regs slow cleanup of hazardous waste. *Consulting Engineer* (July 1983).
7. USEPA. 1982. *Guidance Manual for Hazardous Waste Incinerator Permits.* Washington, DC, Government Printing Office.
8. USEPA. January 1977. *Municipal incinerator enforcement manual.* Washington, DC: Government Printing Office, EPA 340/1-76-013.

RINGLEMANN SMOKE CHART

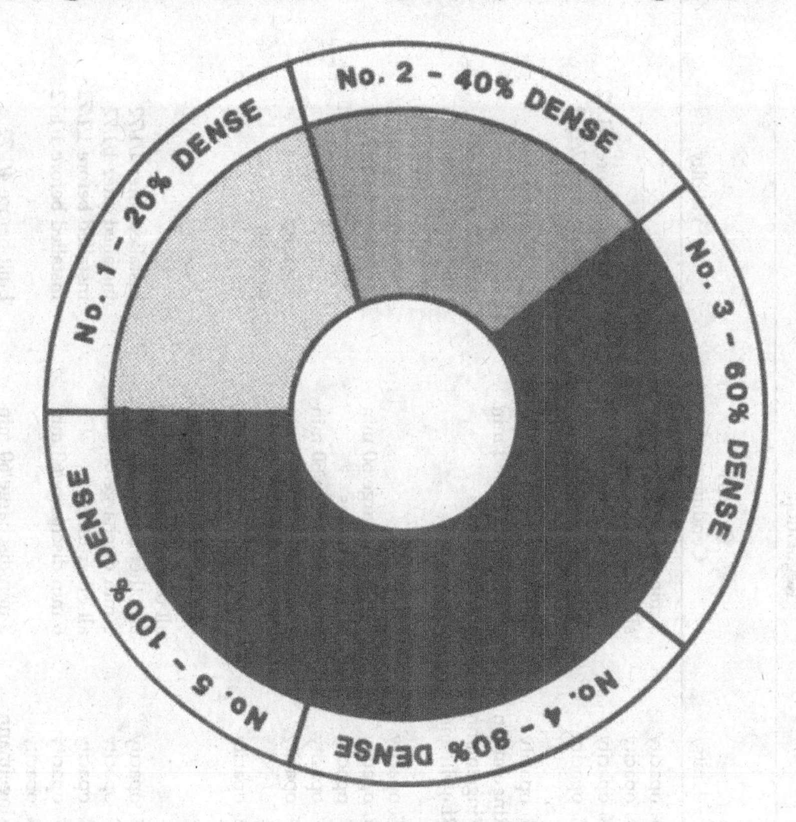

INSTRUCTIONS

1. Hold chart at arm's length and view smoke through circle provided.
2. Observer should not be less than 100 ft. nor more than ¼ mile from the stack.
3. Line of observation should be at right angles to the direction the smoke travels and viewed in the same light.
4. Do not try to observe smoke into direct sun or setting sun. Sun should always be overhead.
5. Match smoke with corresponding color on chart—note density number and time.

Figure 2–1 Opacity measurement.

Table 2–4 Opacity Regulations for New and Existing Commercial and Industrial Incinerators

	Regulation				Equivalent
States	Value	Units	Process Conditions	Validity	Common Regulation (% opacity)
1 Alabama	60	% opacity	3 min discharge/60 min		60
	20	% opacity	all other times		20
2 Alaska	40	% opacity		installed before 7/1/72	40
	20	% opacity		installed after 7/1/72	20
3 Arizona	exempt		0.5 min discharge/60 min		exempt
	20	% opacity	all other times		20
4 Arkansas	No. 3	Ringelmann	5 min discharge/60 min		60
	No. 1	Ringelmann	all other times	built after 7/30/73	20
	No. 2	Ringelmann		built before 7/30/73	40
5 California					
6 Colorado	20	% opacity			20
7 Connecticut	40	% opacity	5 min discharge/60 min		40
	20	% opacity	all other times		20
8 Delaware	20	% opacity	3 min discharge/60 min		20
9 District of Columbia	20	% opacity	2 min discharge/60 min	existing	20
	prohibited		all other times	existing	prohibited
10 Florida	20	% opacity	≤50 TPD, 3 min discharge/60 min		20
	prohibited		all other times		prohibited
11 Georgia	20	% opacity	all other times	installed after 1/1/72	20
	40	% opacity	6 min discharge/60 min	installed after 1/1/72	40
	40	% opacity	all other times	installed before 1/1/72	40
	60	% opacity	6 min discharge/60 min	installed before 1/1/72	60
12 Hawaii	40	% opacity			40
13 Idaho	No. 2	Ringelmann	3 min discharge/60 min	built before 4/1/72	40
	No. 1	Ringelmann	3 min discharge/60 min	built after 4/1/72	20

14 Illinois	30	% opacity	all other times		30
	30–40	% opacity	3 min discharge/60 min		30–60
15 Indiana	40	% opacity	15 min discharge/60 min		40
16 Iowa	40	% opacity			40
	60	% opacity	3 min discharge/60 min during breakdowns, etc.		60
17 Kansas	20	% opacity			20
18 Kentucky	20	% opacity			20
19 Louisiana	No. 1	Ringelmann	all other times		20
	>No. 1	Ringelmann	4 min discharge/60 min		>20
20 Maine	No. 1	Ringelmann			20
21 Maryland					
22 Massachusetts	No. 1	Ringelmann			20
23 Michigan	40	% opacity	3 min discharge/60 min		40
	20	% opacity	all other times		20
24 Minnesota	20	% opacity			20
25 Mississippi	40	% opacity			40
26 Missouri	No. 1	Ringelmann		built after 2/10/72	20
	No. 2	Ringelmann		built before 2/10/72	40
27 Montana	10	% opacity			10
28 Nebraska	20	% opacity			20
29 Nevada	20	% opacity	1 min discharge/60 min		20
30 New Hampshire	No. 1	Ringelmann	3 min discharge/60 min		20
31 New Jersey	No. 2	Ringelmann	3 consecutive minutes		40
	No. 1	Ringelmann	all other times		20
32 New Mexico	No. 1	Ringelmann	2 min discharge/60 min		20
33 New York (state)	40	% opacity		built before 1/26/67	40
(state)	20	% opacity		built after 1/26/67	20
(city)	No. 1	Ringelmann	3 min discharge/60 min		20
34 North Carolina					
35 North Dakota	No. 3	Ringelmann	4 min discharge/60 min		60
	No. 1	Ringelmann	all other times		20

Table 2–4 Opacity Regulations for New and Existing Commercial and Industrial Incinerators (*Continued*)

	Regulation				Equivalent
States	Value	Units	Process Conditions	Validity	Common Regulation (% opacity)
36 Ohio	60	% opacity	3 min discharge/60 min		60
	20	% opacity	all other times	20	
37 Oklahoma	No. 1	Ringelmann	all other times	20	
	No. 3	Ringelmann	5 min discharge/60 min		60
38 Oregon	40	% opacity	3 min discharge/60 min	built before 5/1/70	40
	20	% opacity	3 min discharge/60 min	built after 6/1/70	20
39 Pennsylvania	20	% opacity	3 min discharge/60 min		20
40 Puerta Rica	20	% opacity	all other times		20
	60	% opacity	6 min discharge/60 min		60
41 Rhode Island	20	% opacity	3 min discharge/60 min		20
42 South Carolina	No. 1	Ringelmann	3 min discharge/60 min		20
43 South Dakota	20	% opacity	all other times		20
	60	% opacity	3 min discharge/60 min		60
44 Tennessee	20	% opacity	5 min discharge/60 min		20
45 Texas	30	% opacity	5 min average	built before 1/31/72	30
	20	% opacity	5 min average	built after 1/31/72	20
46 Utah	No. 1	Ringelmann			20
47 Vermont	40	% opacity	6 min discharge/60 min	built before 4/30/70	40
	20	% opacity	6 min discharge/60 min	built after 4/30/70	20
48 Virginia	20	% opacity			20
49 Washington	20	% opacity	3 min discharge/60 min		20
	>20	% opacity	15 min/8 hr		>20
50 West Virginia	No. 1	Ringelmann			20
51 Wisconsin	20	% opacity		built after 4/1/72	20
52 Wyoming	20	% opacity			20

3

National Ambient Air Quality Standards

As part of the Clean Air Act, the federal government has established National Ambient Air Quality Standards (NAAQS) which attempt to define the desired, or permissible, maximum levels of pollutants in the air throughout the country. The pollutants which are used to define the quality of air are termed the *criteria pollutants*, namely, particulate matter, sulfur oxides, carbon monoxide, nitrogen dioxide, lead, and photochemical oxidants (ozone), as listed in Table 3-1. Additional pollutants are regulated under the Clean Air Act; however, clean air is defined in terms of only these six criteria pollutants. The NAAQS for criteria pollutants is listed in Table 3-2 along with the standard for nonmethane hydrocarbons.

Clean Air Standards

The NAAQS establishes clean air standards for minimal (Class I), moderate (Class II), and extensive (Class III) growth areas. In conjunction with the NAAQS, geographical locations are designated as either attainment or nonattainment areas:

Table 3–1 Regulated Pollutants

Criteria Pollutants	Noncriteria Pollutants
Carbon Monoxide	Asbestos
Nitrogen Oxides	Beryllium
Sulfur Dioxide	Mercury
Particulate Matter	Fluorides
Ozone	Vinyl Chloride
(regulate volatile organic compounds)	Sulfuric Acid Mist
Lead	Hydrogen Sulfide
	Total Reduced Sulfur
	Reduced Sulfur Compounds

Source: "Prevention of Significant Deterioration" In *US EPA Workshop Manual* (Washington, DC: Government Printing Office, October 1980).

Table 3–2 National Ambient Air Quality Standards

Pollutant	Averaging Time	Standard Levels Primary	Secondary
Particulate Matter	Annual (geometric mean)	75 μg/m³	60 μg/m³
	24 hr*	260 μg/m³	150 μg/m³
Sulfur Oxides	Annual (arithmetic mean)	80 μg/m³ (.03 ppm)	—
	24 hr*	365 μg/m³ (.14 ppm)	—
	3 hr*		1,300 μg/m³ (0.5 ppm)
Carbon Monoxide	8 hr*	10 mg/m³ (9 ppm)	10 mg/m³ (9 ppm)
	1 hr*	40 mg/m³ (35 ppm)	40 mg/m³ (35 ppm)
Nitrogen Dioxide	Annual (arithmetic mean)	100 μg/m³) (.05 ppm)	100 μg/m³ (.05 ppm)
Ozone	1 hr*	240 μg/m³ (.12 ppm)	240 μg/m³ (.12 ppm)
Hydrocarbons (nonmethane)	3 hr (6 to 9 A.M.)	160 μg/m³ (.24 ppm)	160 μg/m³ (.24 ppm)
Lead	3 mo	1.5 μg/m³	1.5 μg/m³

*Not to be exceeded more than once a year.

Primary Standard: One that is required to protect the public health with an adequate margin of safety.

Secondary Standard: One that adequately protects the public welfare. Public welfare is defined as including, but not limited to, effects on soils, water, crops, vegetation, man-made materials, animals, wildlife, weather visibility, climate, economic values, and personal comfort and well-being.

Source: U.S. EPA, Ambient Air Quality Standards, 40 CFR 50, 1981.

- Attainment area: Air in this geographical location is at present considered clean; that is, within the definition of clean air in the NAAQS for the designated growth area classification. The air quality for all criteria pollutants is acceptable.
- Nonattainment area: Air quality in a nonattainment area is below the quality established in the NAAQS for the designated growth area classification. An area may be classified as nonattainment for one or more of the criteria pollutants depending upon its concentration in the atmosphere compared to the levels listed in Table 3-2. The same

location can be an attainment area for sulfur oxide and particulate matter emissions but, at the same time, be a nonattainment area for carbon monoxide, nitrogen dioxide, and ozone.

The United States has been divided into 247 separate Air Quality Control Regions (AQCRs) in accordance with the provisions of the Clean Air Act. The status of each of these AQCRs has been determined (attainment or nonattainment) for each of the criteria pollutants. This status evaluation changes with time in response to the introduction of new sources in the region, more effective control of existing sources, or the closing of existing sources.

Prevention of Significant Deterioration

A set of regulations has been established by the federal government to help achieve the air quality standards designated by the Clean Air Act. These are known as the Prevention of Significant Deterioration regulations (PSD). A significant feature of the PSD is the New Source Review (NSR). Under the PSD a significant source of air pollution must be analyzed, under an NSR procedure, to determine the effect of that source on the surrounding geographical area. In addition, methods are stipulated for a systematic review of the new source to determine the adequacy of its proposed air pollution control equipment. This review addresses the question of Best Available Control Technology (BACT).

Implementation of NSR and BACT procedures is required, as noted above, for significant sources of air pollution. The PSD regulations define a significant source as either of the following

- Any source which emits or has the potential to emit over 250 tons/year of any regulated pollutant into the atmosphere. Pollutants are any of those pollutants which are listed in Table 3-3.
- Any one of the 28 source categories listed in Table 3-4 which emits or has the potential to discharge more than 100 tons/year of any regulated pollutant into the atmosphere.

If a source emits any regulated pollutant in an amount greater than the 100-or 250-ton threshold but has specific pollutant emissions in excess of the "significant emission" rate value for any pollutant (see Table 3-3), a BACT review is necessary to limit the emission potential of that particular pollutant. For instance, if a plant which has sources subject to PSD generates a total of 65 tons/year of the pollutants listed in Table 3-3, a new source review does not have be performed. If, however, sulfur dioxide makes up over 40 tons/year of the 65 ton/year total, a BACT review must be effected for sulfur dioxide control equipment.

These pollutants are measured at the stack discharge. The term *potential* refers to the toal amount of pollutants which may be discharged by

Table 3–3 Significant Emission Rates

Pollutant	Emission Rate tons/year
Carbon Monoxide	100.
Nitrogen Oxides as NO_2	40.
Sulfur Dioxide	40.
Total Suspended Particulate	25.
Ozone (Volatile Organic Compounds)	40.
Lead	0.6
Asbestos	0.007
Beryllium	0.0004
Mercury	0.1
Vinyl Chloride	1.0
Fluoride	3.0
Sulfuric Acid Mist	7.0
Total Reduced Sulfur (including H_2S)	10.0
Reduced Sulfur (including H_2S)	10.0
Hydrogen Sulfide	10.0

Source: "Prevention of Significant Deterioration," *U.S. EPA Workshop Manual* (Washington, DC: Government Printing Office, October, 1980).

the process in question. For instance, although a unit may be in operation only 30 weeks a year, it may have the ability to run for 52 weeks a year. The potential emissions, therefore, refer to, in this illustration, 52 weeks per year operation although the process may in actuality operate only a fraction of the year. There is an important exception to this rating. If the hours of operation can be limited by a federerally enforceable permit the potential discharge can be considered for the proposed operating schedule.

The New Source Review Process

Figure 3-1 illustrates the new source review (NSR) process. With reference to this chart:

If an installation does not consittute a major new (or modified) source, it is not subject to federal PSD review. (A major modification is one which results in the emission of a pollutant in excess of its "significant emission rate" value.) The installation, however, must comply with applicable emissions regulations of the state in which it is located. Some states have adopted modified provisions of the National Ambient Air Quality Standards (NAAQS) and may require an analysis, or computer modeling, more stringent than is required under NAAQS. (Note that

Table 3–4 Major Stationary Sources of Air Pollution

Coal Cleaning Plants (with thermal dryers)
Kraft Pulp Mills
Portland Cement Plants
Primary Zinc Smelters
Iron and Steel Mills
Primary Aluminum Ore Reduction Plants
Primary Copper Smelters
Municipal Incinerators Capable of Charging Over 250 tons of Refuse per Day
Hydrofluoric Acid Plants
Sulfuric Acid Plants
Nitric Acid Plants
Petroleum Refineries
Lime Plants
Phosphate Rock Processing Plants
Coke Oven Batteries
Carbon Black Plants (furnace process)
Primary Lead Smelters
Fuel Conversion Plants
Sintering Plants
Secondary Metal Production Facilities
Chemical Process Plants
Fossil-fuel Boilers of more than 250 million Btu/hr heat input
Petroleum Storage and Transfer Facilities with a capacity exceeding 300,000 barrels
Taconite Ore Processing Facilities
Glass Fiber Processing Plants
Charcoal Production Facilities

models or regulations designated by any state can never be less stringent than those designated by the federal government.) A public hearing may or may not be necessary and a decision to hold a hearing is governed by state statute.

If a source is a major new or modified source, as defined previously, a determination must first be made as to its location. If it is not located in a nonattainment area the PSD rules require that an assessment of the BACT be applied to each pollutant emission in excess of the significant emissions rate value. The BACT analysis evaluates the state-of-the-art in air pollution control equipment for the application in question and considers the economic and energy impacts of that equipment.

A screening model is applied for a BACT analysis. This is a relatively simple mathematical modeling technique, for use with a computer, which is used to determine the order of magnitude of emissions resulting from the BACT analysis on the NAAQS. Ambient air monitoring is not required.

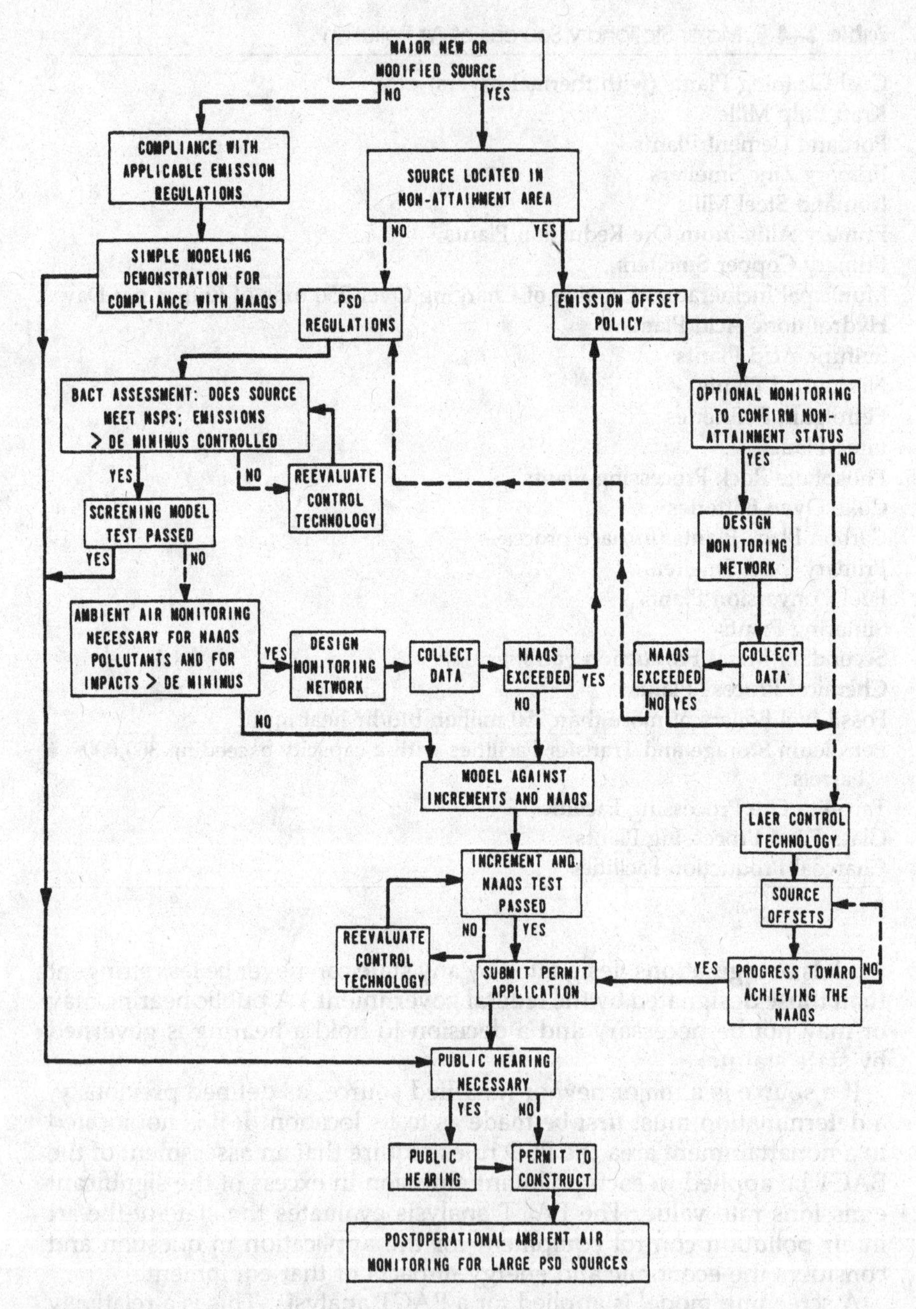

Figure 3–1 Principal steps in the new source review process.

If the result of the screening model is a concentration less than 50% of that allowable under the NAAQS, further source study is not necessary and the permitting question can go to a public hearing. If the screening model indicates that a discharge equivalent to greater than 50% of the NAAQS will occur then further analysis must be performed. Refined ambient air modeling must be used in this case; for example, a complete, fairly complex computer modeling procedure is necessary. This procedure utilizes inputs from the anticipated source (pollutants discharged, stack discharge temperature, stack gas velocity, etc.), meteorological data (prevailing winds, latitude, storm frequency, mean temperature, precipitation, etc.) and receptor data (topology, i.e., presence of hills or mountains and their height, water surfaces, location of other structures, etc.). If this modeling procedure indicates that the NAAQS will not be met, additional emissions reductions must be instituted. If the modeling results indicate that the NAAQS will be satisfied, the way is open for a public hearing.

The above discussion concerned a source located in an attainment area. If the new (or modified) source were located in a nonattainment area, procedures of the emissions offset policy (EOP) must be followed for those pollutants whose impacts exceed significant ambient air quality levels (see Table 3-3). The EOP requires the following:

- Source emissions must be controlled to the greatest degree possible.
- More than just offsetting emissions reduction must be secured for existing sources on the same site.
- Progress must be made for achieving the NAAQS.

After the EOP is satisfied the adequacy of the new source with respect to emissions can be demonstrated by a modeling procedure or by implementation of the lowest achievable emissions rate (LAER) review. This review procedure is similar to the BACT review. The BACT review combines state-of-the-art technology with economic and energy considerations. The LAER is much more severe. It is a BACT analysis, but it does not consider the economic or energy impact of the proposed control system. Table 3-5 lists control equipment for municipal incinerators generally found applicable to BACT and LAER procedures. An example of the results of these two procedures is sulfur dioxide and sulfuric acid mist control. In both of these cases when considering economic and energy factors (BACT), no control system has practicality. On the other hand, when considering only the technological aspect of control (LAER) scrubbers are found to be effective, and desirable.

If the modeling procedure is utilized at the conclusion of the EOP analysis and if it is found that applicable significant ambient air quality levels are exceeded, the LAER review must be instituted. Otherwise, the source can be considered as if it were located in an attainment area.

Table 3–5 State of the Art Control Systems, Refuse Incinerators

Pollutant	BACT	LAER
Particulates	ESP	ESP
Sulfur Dioxide	none	DRY OR WET SCRUBBER
Nitrogen Oxides as NO_2	none	none
Carbon Monoxide	CC[a]	CC
Hydrocarbons	CC	CC
Hydrochloric Acid	none	DRY OR WET SRUBBER
Fluorides	none	DRY OR WET SCRUBBER
Lead	ESP[b]	ESP[b]
Mercury	none	none
Berryllium	ESP[b]	ESP[b]
Sulfuric Acid Mist	none	DRY SCRUBBER
Tetrachlorodibenzo-p-dioxins	none	none
Polynuclear Aromatics	none	none
Polychlorinated Biphenyls	none	none
Asbestos	none	none
Hydrogen Sulfide	none	none
Vinyl Chloride	none	none
Reduced Sulfur	none	none

[a]standard modern combustion controls
[b]controlled concurrently with particulate

Source: W. O'Connell, G. Stotler, and R. Clark, "Emissions and Emission Control in Modern Municipal Incinerators, *Proceedings* of the 1982 National Waste Processing Conference, ASME (New York: ASME 1982).

Application of the LAER procedure must, in all cases, demonstrate that there is progress toward meeting the NAAQS before a public hearing can be advertised.

An agency review (by the Environmental Protection Agency [EPA], its state representative, or the cognizant state agency) of the above procedure must prove successful before a public hearing is called. When a permit is granted it is done so only after a public hearing has been held and a designated comment period has elapsed.

New Source Review Implementation

The NSR procedure is exceedingly complex. It can cost hundreds of thousands or even millions of dollars and will add at least 1, and perhaps 2 to 4 years to the start of construction. In addition, this review can open up the entire operation of a firm to public scrutiny, which in turn can provide public detractors with ready-made delay and obstructionist tactics, further impeding progress of the project.

The purpose of this review is laudable and it is a workable procedure. Table 3-6 illustrates its application: in 1980 80% of the power plants, 87% of the pulp and paper mills, and 83% of the municipal incinerators in the United States were operating in compliance with the Clean Air Act and its derivative procedures. This review, however, should be applied only when necessary. There will be situations where process parameters can be changed in order to be under the threshold at which a PSD review is triggered. Decreasing the temperature of destruction by decreasing the excess air injected into an incinerator will decrease the generation of nitrogen oxides. If this pollutant is crucial in establishment of a PSD requirement, such process modifications should be investigated. Similarly, modifications of operating parameters can be implemented for other pollutants.

Pollutant Standards Index

The Pollutant Standards Index, (PSI), is a measure of overall air quality based on the value of the concentrations of five of the six criteria air pollutants (sulfur dioxide, nitrogen dioxide, carbon monoxide, ozone, and total suspended particulate matter). It is a single number that can range from 0 to 500, 0 being the highest and 500 the lowest ambient air quality. Air quality is considered "good" with a PSI less than 50, "moderate" between 50 and 99, "unhealthful" between 100 and 199, "very

Table 3–6 Compliance Status of Major Air Pollution Sources, 1980

		In Compliance With Emissions Limitations	
Industry	Number of Total Sources	Number	Percent
Coal Cleaning	409	395	97
Asphalt Concrete	2862	2752	96
Sulfuric Acid	262	246	94
Phosphatic Fertilizers	69	62	90
Portland Cement	200	176	88
Gray Iron	433	381	88
Pulp and Paper	475	417	87
Municipal Incinerators	72	60	83
Power Plants	700	559	80
Petroleum Refineries	214	170	79
Aluminum Reduction	49	37	76
Iron and Steel	204	110	54
Primary Smelters	28	13	46

Source: Council on Environmental Quality, *Environmental Quality*, The Eleventh Annual Report (Washington, DC: Government Printing Office).

Table 3–7 Ranking of 40 Standard Metropolitan Statistical Areas Using the psi, 1976–1980

Severity Level	SMSA	Number of Days Per Year psi > 100	psi > 200
Over 150 Days	Los Angeles	242	118
	New York	224	51
	Pittsburgh	168	31
	San Bernardino	167	88
100–150 Days	Cleveland	145	35
	St. Louis	136	29
	Chicago	124	21
	Louisville	19	12
50–99 Days	Washington, DC	97	8
	Phoenix	84	10
	Philadelphia	82	9
	Seattle	82	4
	Salt Lake City	81	18
	Birmingham	75	19
	Portland	75	3
	Houston	69	16
	Detroit	65	4
	Jersey City	65	4
	Baltimore	60	12
	San Diego	52	6
25–49 Days	Cincinnati	45	2
	Dayton	45	2
	East Chicago	36	8
	Indianapolis	36	2
	Milwaukee	33	6
	Buffalo	31	5
	San Francisco	30	1
	Kansas City	29	6
	Memphis	28	2
	Sacramento	28	2
	Allentown	27	1
0–24 Days	Toledo	24	2
	Dallas	22	1
	Tampa	12	1
	Akron	10	0
	Norfolk	9	0
	Syracuse	9	1
	Rochester	6	0
	Grand Rapids	5	0

Source: Council on Environmental Quality, *Environmental Quality—1980*, The Eleventh Annual Report (Council on Environmental Quality, December 1980), 154.

unhealthful" between 200 and 299 and "hazardous" at 300 or above. A PSI will exceed the value of 100 when any one of these five criteria pollutants exceeds the NAAQS for that geographical area.

Table 3-7 lists PSI occurrences over 100 and over 200 for the 40 largest metropolitan areas in the country. During the 5-year sampling period over 40 million people lived in a part of the country where the air was classified as "unhealthy" for more than half of the year. In all but four of the areas listed, which represents where over 75% of all Americans live and work, the air was classified "very unhealthful" at least once in each of the years of the sampling period. The contribution of incineration to these figures is relatively small, with automobile emissions the largest single cause of bad air quality in most of these areas.

References and Bibliography

1. Ambient Air Quality Standards, 40 CFR 50, Government Printing Office, Washington, DC, 1981.
2. Matey, J. The clean air act, success amid a perception of failure. *Pollution Engineering* (November 1982).
3. Commission on Natural Resources. 1981. *On Prevention of Significant Deterioration of Air Quality*, Washington, DC: Government Printing Office, 1981.
4. Sweitzer, T. Understanding the ambient air monitoring regulations. *Pollution Engineering* (February 1980).
5. Gunther, C. Resource recovery and the clean air act. *Waste Age*. (May 1981).
6. USEPA. 1979. *A review of standards of performance for new stationary sources, incinerators, EPA 450/3-79-009*, Washington, D.C. Government Printing Office.
7. USEPA. 1978. *A review of standards of performanace for new stationary sources, sewage sludge incinerators*. EPA 450/3-79-010. Washington, DC. Government Printing Office.
8. Axell, K., T. Devitt, and N. Kulujian. 1975. *Inspection manual for enforcement of new source performance standards, municipal incinerators, EPA 340/1-75-003*. Washington, DC. Government Printing Office.

4

Particulate Pollutants

Particulate matter is both solid and liquid. The term comprises a complex category of materials, also termed *aerosols*, that inhabit the atmosphere. The size range of particulates which is of interest varies from just over that of large individual molecules, 0.1μ (microns), to 500μ in diameter, where 1 million μ equal 1 meter(m). Particles above 10μ in diameter can be seen with the unaided eye. With a conventional microscope particles as small as 0.5μ can be identified. Electron microscopes are necessary to identify individual particles down to approximately 0.05μ in diameter.

Some natural sources of particulate emissions are listed in Table 4–1. Primary pollutants are those that are actually discharged from a source, whereas secondary pollutants are not discharged but are formed external to the source. Secondary pollutants may be formed by the action of

Table 4–1 Origins of Ambient Urban Aerosols

Source	Estimated Contribution (Percent of TSP)		
	Washington, D.C.	Pasadena, California	Chicago Illinois
Primary:			
Crystal Dust	24.	11.4	18.
Limestone (Cement)	4.1	1.7	3.2
Sea Salt	0.9	1.3	–
Coal Burning	6.3	–	6.4
Residual Oil Combustion	0.6	0.1	1.4
Refuse Incineration	1.4	–	–
Motor Vehicles	7.1	13.5	2.8
Steel Processing	–	–	3.9
Other Industry	–	7.3	–
Secondary:			
NH_4^+	1.4	–	–
SO_4^-	14.	>10.	11.5
NO_3^-	3.2	0.1	5.3
Volatile Carbon	–	22.	–

Source: Committee on Particulate Control Technology, Commission on Natural Resources, National Research Council, National Academy of Sciences, *Controlling Airborne Particulates* (Washington, DC: National Academy of Sciences, 1980), 36.

sunlight on a primary pollutant or may be generated by the action of one pollutant upon another.

Crystal dust, the major aerosol discovered in the air over Washington, DC and Chicago, comprises soil and dust blown by the wind. A major component of this source is unpaved roads. The relatively high amount of limestone dust found in the air environment results, in large measure, from construction activities (mixing and pouring concrete, removing concrete forms, demolition, sandblasting, etc.), from agricultural liming, erosion of streets and buildings, and cement and lime plants.

Refuse incineration accounts for a relatively small amount of the total identifiable particulate sources, as indicated in Table 4–1. There are refuse incinerators within the Washington, DC and Chicago areas, but their contribution to the total particulate load appears negligible.

Table 4–2 lists sources of natural aerosols found in the atmosphere over the United States on a daily basis. A list of man-made, or anthropogenic sources of aerosol pollution is presented in Table 4–3. The total particulate loading due to industrial and agricultural activity in this country is only 7% of the aerosol loading produced by nature. (The maximum estimated loading from anthropogenic sources, 700,000 tons, divided by the estimated 10 million tons from natural sources.)

These tables would appear to indicate that the aerosols generated by man are not very significant, that they are, in total, less than that produced by forest fires, a natural occurrence. Quantity, however, is not the only parameter of interest when evaluating air quality or emissions

Table 4–2 Sources of Natural Aerosols in the Atmosphere

Source	Estimated Aerosol Production Tons Per Day
Primary:	
Dust Rise By Wind	20,000–1,000,00
Sea Spray	3,000,000
Forest Fires (intermittent)	400,000
Volcanic Dust (intermittent)	10,000
Extraterrestrial (meteorite dust)	50–550
Secondary:	
Vegetation: Hydrocarbons	500,000–3,000,000
Sulfur Cycle: SO_4 – –	100,000–1,000,000
Nitrogen Cycle: NO_3 –	1,000,000
NH_4 +	700,000
Volcanoes: Volatile SO_2 and H_2S (intermittent)	1,000
Maximal Total (approximate)	10,000,000

Source: National Academy of Sciences, *Biologic Effects of Atmospheric Pollutants* (Washington, DC: National Academy of Sciences, 1977), 37.

Table 4–3 Some Sources of Anthropogenic Aerosols in the Atmosphere

Source	Estimated Aerosol Production Tons Per Day
Primary:	
Combustion and Industry	100,000–300,000
Dust Rise by Cultivation	100–1,000
Secondary:	
Hydrocarbon Vapors	7,000
Sulfates (SO_2, H_2S to SO_4− −)	300,000
Nitrates (NO_x to NO_3−)	60,000
Maximal Total (approximate)	700,000

Source: National Academy of Sciences, *Biological Effects of Atmospheric Pollutants* (Washington, DC: National Academy of Sciences, 1979).

significance. Particle size, their physical and chemical nature, are all factors important in the evaluation of the pollution source.

In general, man-made particles greater than 10μ mean diameter come from mechanical processes such as erosion, grinding, and spraying. Those from one to 10μ originate partly from mechanical processes and also include industrial dusts and ash. Among particles in the size range 0.1μ to 1μ, sulfates and the products of combustion begin to predominate, along with other aerosols formed by chemical and photochemical oxidation in the air. Particles larger than 2μ to 5μ settle on surfaces and constitute "dustiness."

Table 4–4 lists particle size ranges for various natural and man-made aerosols.

Particulate Properties

Chemical properties of particles will vary with the nature of the particles, as will biological properties. The physical properties of a particle will be as diverse as its chemical and biological properties; however, three general physical properties can reasonably be said to apply to all particulate matter. These three properties all are concerned with the interface between the particle and its surroundings. They can be considered as surface, motion, and optical properties:

Surface Properties

The three surface properties of interest are sorbtion, nucleation, and adhesion.

Sorbtion. Sorbtion can be explained by considering the impact of individual molecules on a particle surface. If the impact is perfectly elastic

Table 4–4 Particle Size Ranges for Aerosols

Substance	Approximate Range of Particle Mean Diameter (microns)
Rain Drops	500 — 5,000
Natural Mist (water vapor)	60 — 500
Natural Fog and Clouds (water vapor)	2 — 60
Stoker Fly Ash	10 — 800
Pulverized Coal Fly Ash	1 — 50
Foundry Dusts	1 — 1,000
Cement Dusts	3 — 100
Metallurgical Dust	0.5 — 100
Pollens	10 — 100
Ground Talc	0.5 — 50
Bacteria	0.3 — 35
Plant Spores	10 — 35
Sulfur Trioxide Mist	0.3 — 3
Insecticide Dusts	0.5 — 10
Pigments (paint)	0.1 — 5
Ammonium Chloride Fume	0.1 — 3
Alkali Fume	0.1 — 5
Oil Smoke	0.1 — 1.0
Metallurgical Fume	0.01 — 2.2
Resin Smoke	0.01 — 1.0
Tobacco Smoke	0.01 — 1.0
Normal Impurities in Quiet Air	0.01 — 1.0
Carbon Black	0.01 — 0.3
Colloidal Silica	0.02 — 0.05
Zinc Oxide Fume	0.01 — 0.5
Magnesium Oxide Fume	0.01 — 0.5
Sea Salt Nuclei	0.03 — 0.5
Combustion Nuclei	0.01 — 0.1
Virus and Protein	0.003 — 0.05
Gas Molecules (diameter)	0.0001 — 0.0006
Dust Damaging to the Lung (silicosis)	0.5 — 5
Human Hair	35 — 200
Red Blood Cells (adults)	7.5

Source: American Industrial Hygiene Association, *Air Pollution Manual*, Part 1 (Evaluation), 2d ed. (Westmont, NJ: AIHA, 19), 17.

then the molecule will rebound from the surface instantaneously. If rebound from the surface is delayed or if the velocity of rebound is significantly smaller than that of impact there will be a local accumulation of molecules on or near the particle surface. If the delay is great, a substantial fraction of the surface may be covered. This phenomenon is called adsorption. If a delay is caused by a chemical interaction between the surface and the impacting molecule the process is known as chemisorbtion. Absorbtion is that phenomenon where the molecule is dissolved into, or within, the particle.

Nucleation. Nucleation refers to that phenomenon where foreign material collects on a particle developing larger, "nuclear" particles in a population. For instance, a vapor present at or near its equilibrium vapor pressure may accumulate a deep sorbed layer on a particle, which then takes on the character of a true liquid or solid. If the vapor is supersaturated, a droplet or crystal may grow by further condensation on the sorbed layer. The net result is nucleation.

A complete sorbed layer on a particle surface causes that particle to behave as a drop of the larger particle diameter. Particles are always present in the atmosphere with sorbtion and nucleation occurring constantly.

Adhesion. Adhesion is a physical property associated with particle size. Solid particles with mean diameters less than 1μ and liquid particles regardless of size normally adhere when they collide with each other or with a larger surface. Other factors being equal, rebound becomes increasingly probable with increasing particle size. Conversely, adhesion decreases in probability as the particle size increases.

Motion

A property common to all particles, regardless of composition, is particulate motion. Particles with mean diameters less than 0.1μ undergo large random motions caused by collision with individual molecules, the phenomenon known as Brownian motion. Larger size particles, above 1μ in diameter, have relatively high settling velocities, and their motions can vary significantly from the motion of the air on which they are carried. Particles in the size range of 0.1μ to 1μ have finite settling velocities, but these velocities are small compared with air motion. Particles larger than 5μ to 10μ are removed from the air by gravity and other inertial or settling processes. The size distribution of a freshly generated aerosol may contain particles as large as 500μ but these larger particles settle rapidly. Particles in the size range of approximately 5μ to 10μ require significant periods of time for natural elimination from their distribution.

The net result of the Brownian motion and settling mechanisms is a fairly consistent rate of fallout of larger particles from the air and maintenance of smaller particles airborne. This tends to produce a fairly uniform particle size distribution.

Optical Properties

It is through their optical effects that particles are normally perceived in the atmosphere. Particles in the range of 0.1μ to 1μ exhibit properties showing a transition between two extremes.

Below 0.1μ, particles are sufficiently small compared to the wavelength of light to obey approximately the same laws of light scattering as do molecules, termed *Raleigh scattering*, a function of sixth power of the particle diameter. This mechanism is relatively inconsequential in its effect upon visibility. Particles very much larger than 1μ are so much larger than the wavelength of visible light that they obey the laws of macroscopic objects, intercepting or scattering light roughly in proportion to their cross-sectional area. Figure 4–1 illustrates the effect of light scattering as a function of particle size. Light scattering, or the extinction of light, is greatest with particles in the range of 0.1 to 1 micron.

Smoke

Smoke is the most easily identifiable pollutant from a source. It is, technically, a suspension of solid or liquid particulate matter in a gaseous discharge. The particles range from fractions of a micron to over 50μ in

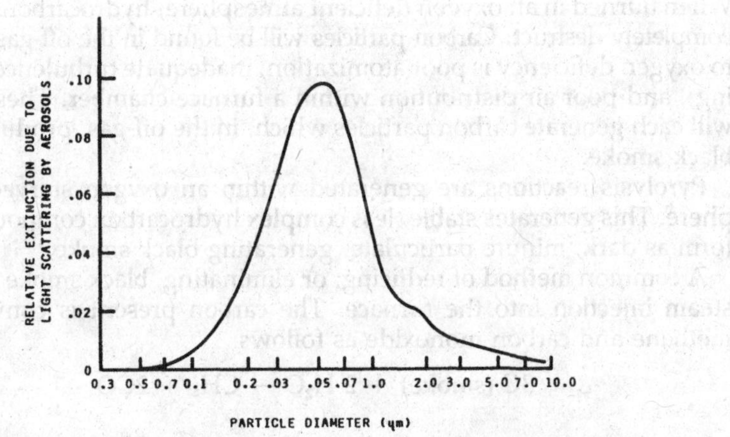

Figure 4–1 Light scattering as a function of particle diameter.

diameter. The visibility of smoke is related to the quantity of particles present, rather than the weight of the particulate matter. The weight of particulate emissions is therefore not necessarily indicative of the density of the emission. For instance, a weight of so many grains of emissions per cubic foot of gas is not directly related to the opacity of the discharge. Neither is the color of a discharge related to opacity, or smoke density. Smoke can be either black or can appear nonblack, termed *white smoke*:

White Smoke

The formation of white or other opaque, nonblack smoke is usually due to insufficient furnace temperatures when burning carbonaceous materials. Hydrocarbons will be heated to a level where evaporation and/or cracking will occur within the furnace when white smoke is produced. The temperatures will not be high enough to produce complete combustion of these hydrocarbons. With a stack temperature in the range of 300°F to 500°F, many of these hydrocarbons will condense to liquid aerosols and with the solid particulate present these will appear as nonblack smoke.

An increase in the furnace/stack temperatures and increased turbulence is one method of controlling white smoke. Turbulence helps insure uniformity of this higher temperature within the off-gas flow.

Excessive airflow may provide excessive cooling, and an evaluation of reducing white smoke discharges would include investigating the air quantity introduced into the furnace. Inorganics in the exit gas may also produce a nonblack smoke discharge. For instance, sulfur and sulfur compounds will appear yellow in a discharge, calcium and silicon oxides in the discharge will appear light to dark brown.

Black Smoke

When burned in an oxygen deficient atmosphere, hydrocarbons will not completely destruct. Carbon particles will be found in the off-gas. Related to oxygen deficiency is poor atomization, inadequate turbulence (or mixing), and poor air distribution within a furnace chamber. These factors will each generate carbon particles which, in the off-gas, produces dark, black smoke.

Pyrolysis reactions are generated within an oxygen starved atmosphere. This generates stable, less complex hydrocarbon compounds that form as dark, minute particulate, generating black smoke.

A common method of reducing, or eliminating, black smoke has been steam injection into the furnace. The carbon present is converted to methane and carbon monoxide as follows:

$$3C \text{ (smoke)} + 2\,H_2O \rightarrow CH_4 + 2CO$$

Similar reactions occur with other hydrocarbons present and the methane and carbon monoxide produced burns clean in the heat of the furnace, eliminating the black carbonaceous smoke that would have been produced without steam injection:

$$CH_4 + 2O_2 \rightarrow CO_2 + 2H_2O \text{ (smokeless)}$$

$$2CO + O_2 \rightarrow 2CO_2 \text{ (smokeless)}$$

Steam injection normally requires from 20 to 80 pounds of steam per pound of flue gas.

It should be noted that there is some controversy regarding the effect of steam injection on carbonaceous discharges. Some argue that the steam produces, primarily, good mixing and that the effect of turbulence, or effective mixing with air, is to eliminate the smoke discharge as opposed to methane generation, which may, in fact, be generated in negligible quantities.

Metals Emissions

Metallic particulate discharges into the atmosphere account for less than 1% of the total particulate discharge. However, many of these metals-emitted metals compounds can have severe, toxic effects. It has been found that inhalation of certain metals can have a greater effect on the human body than receiving the same metals damage by digestion.

There are five metals associated with waste combustion which have been shown to have a significant negative effect on human health:

- Lead (Pb): This metal accumulates within the human body. It causes symptoms such as anemia, headaches, sterility, miscarriages, or the birth of handicapped children. Such handicaps, known as lead encephalopathy, include convulsions, coma, blindness, mental retardation, and/or death.
- Nickel: Nickel carbonyl (NiCO) is considered to be a form of nickel which is hazardous to human health. It causes changes in the lung structure, which results in respiratory diseases, including lung cancer. It has been found to be present in tobacco smoke.
- Cadmium (Cd): This element is associated with the incineration of cadmium-containing products such as automobile tires and certain plastics. In the human body it has been found to interfere with the natural processes of zinc and copper metabolism. Cadmium has also been found to cause cardiovascular disease and hypertension.
- Mercury (Hg): Mercury has long been recognized as a pollutant in the water environment. Because of its low boiling point it is also released into the atmosphere when it is present in a combustible, or heated material. Mercurial poisoning in humans is characterized by

blindness, progressive weakening of the muscles, numbness, paralysis, coma, and death. It also leads to severe birth deformities.

- Chromium (Cr): Chromium is normally generated in either the trivalent or hexavalent form, which will be discussed in Chapter 10. The hexavalent form of the metal is considered hazardous whereas the trivalent form is not considered hazardous although it can have toxic effects.

  Hexavalent chromium will cause dermatitis to exposed skin, will irritate the mucous membranes, and may produce pulmonary sensations. It has been found to produce dental erosion, loss of weight, and there is concern that it may be a carcinogen.

Health Effects vs. Particle Size

Particles below 15μ can be retained within the lungs. They are referred to as Inhaled Particulate (IP). Larger size particulate matter will generally be expelled during the breathing process. Of these smaller particles, those under approximately 3μ pose the greatest danger to human health. They have been found to penetrate through to the depths of the lungs where they are more likely to release adsorbed contaminants into the bloodstream.

Table 4–5 lists elements that are found on small particles (below 2μ) generated from the incineration of solid waste and coal. Of the metals discussed above (Pb, Ni, Cd, Hg, and Cr), lead and cadmium are the most prevalent on smaller particles with chromium present to a lesser degree. Nickel, while present in emissions from coal burning plants has not been found in the tabulated examination of small particle refuse incinerator emissions. Mercury has not been discovered on small particles for any of the listed examples.

References and Bibliography

1. National Academy of Sciences. 1980. *Controlling airborne particulates.* Washington, DC: National Academy of Sciences.
2. National Air Pollution Control Administration. 1961. *Control techniques for particulate pollutants.* AP/51. Washington, DC: Government Printing Office.
3. New York City Department of Air Pollution Control. Undated. *Air pollution and smoke control.*
4. Kleinman, M. Identifying and Estimating the Relative Importance of Sources of Airborne Particulates, *Environmental Science and Technology* (January 1980), 14/1: 62–65.
5. Kinsman, R. 1981. Effects of particulate air pollution on asthmatic subjects. *EPA 600/S1-81-033.* Washington, DC: Government Printing Office.

Table 4–5 Fractions of Elements Released From Combustion Sources on Particles Less Than 2 Microns (2μ) in Diameter

Coal-Fired Power Plants With Electrostatic Precipitator (1)		(2)		Refuse Incinerator With Water-Cooled Baffle (3)	
Element	% < 2μ	Element	% < 2μ	Element	% < 2μ
Cr	0.58	Fe	8.8	Ca	14.
Fe	4.5	Se	17.	Sc, Ti	<20.
Al, Na	5.1	Co	25.	Th	24.
Th	5.8	V	27.	Al	28.
Ba	6.4	Na, Al	28.	Mg	<30.
K, Ca	7.5	K, Mg, Ca, Ba, Sc, Ti	29.	Cr	32.
Mn	7.9	Mn, Zn	30.	Fe	38.
V, Ga	9.6	Ni	32.	Co	41.
Se	8.8	Cr	33.	Se	44.
Zn	10.	I, Pb	49.	V	48.
Co, Cl, W	11.	Sb	57.	Mn	55.
As	13.	Br	60.	Ba	58.
Mo	16.	As	63.	Au	72.
Sb	17.			Na	80.
				Cl, W	83.
				Cs	84.
				Br	90.
				Cu	91.
				Zn, Ag	92.
				As, Sb	94.
				Cd, In	95.
				Pb	96.

Sources:
1. J. Ondov, R. Ragaini, "Elemental Particle Size Emissions From Coal Fired Power Plants," *Atmospheric Environment* 12 (1978): 1175–85.
2. E. Gladney, Trace Element Emissions of a Coal Fired Power Plant, Ph.D. diss., Department of Chemistry, University of Maryland, 1974.
3. R. Greenberg, W. Zoller, and G. Gordon, "Composition and Size Distributions of Particles Released in Refuse Incinerators" *Environmental Science ∞ Technology 12 (1978): 573–79.*

6. Bennett, R., and K. Knapp. Characterization of Particulate Emissions from Municipal Wastewater Sludge Incineration, *Environmental Science and Technology* (December 1982) 16/12: 831–836.
7. Dewling, R., R. Manganelli, and G. Baer. Fate and behavior of selected heavy metals in incinerated sludge. *WPCF Journal*. (October 1980) 52/10: 2552–2557.

5

Gaseous Pollutants

The burning process converts fuel to gaseous and solid constituents. The solids produced are either airborne, exiting the furnace in the flue gas, or they are residuals, leaving the furnace as bottom ash. The gaseous discharge will exit the furnace in the flue gas. Flue gas will normally have components classified as either organic or inorganic and they can be innocuous or a danger to health and materials.

Inorganic Gas Discharges

Inorganic gases produced from the burning process normally include carbon dioxide, carbon monoxide, oxides of nitrogen, and when sulfur is present, oxides of sulfur. Carbon dioxide is not considered a pollutant; however, there is concern that excessive quantities of this gas within the atmosphere might produce a "greenhouse effect." This is the mechanism whereby carbon dioxide molecules absorb heat energy and prevent the normal radiation of heat from the Earth, as glass does in a greenhouse. An indication of this effect can be seen by examining the mean temperature of the Earth over the past century. The mean temperature above the surface of the Earth rose approximately 1°F, while at the same time the level of carbon dioxide in the atmosphere rose over 10%.

Carbon monoxide is a danger to human health. It has the ability to pass through the lungs, directly into the bloodstream of an organism, where it destroys the ability of red blood cells to carry oxygen. At an exposure of only 0.10% carbon monoxide in air by volume (1000 ppm) a human being will be comatose in less than 2 hours. The federal government has established a maximum exposure standard for carbon monoxide of 9.0 ppm for an 8-hour average and 13.0 ppm for any hour, where ppm is related to volume.

Nitrogen is an extremely active substance which forms a wide range of compounds with oxygen as listed in Table 5–1. As seen in this table, nitrogen dioxide is the most significant (harmful) pollutant of all of the nitrogen oxides. Federal standards for nitrogen oxides (NO_x) which includes NO and NO_2 expressed as NO_2, are 0.05 ppm maximum on a yearly average and 0.13 ppm maximum for any 24-hour average.

Where NO and NO_2 are present, the calculation for NO_x is as follows:

Table 5–1 Oxides of Nitrogen

Formula	Name	Effects
N_2O	Nitrous Oxide	Inert, not a pollutant (laughing gas). Used as a carrier gas in aerosol containers.
NO	Nitric Oxide	Main product of combustion. Considered harmless by itself. Converts to NO_2. Some indication that it may attach itself to blood hemoglobin and in large concentrations can cause serious oxygen deprivation.
N_2O_3	Dinitrogen Pentoxide	Unstable, rare, not a significant pollutant.
NO_2	Nitrogen Dioxide	Causes significant effects in the Atmosphere, i.e., smog, yellows white fabric, creates plant leaf injury, reduces plant yields.
N_2O_7	Dinitrogen Pentoxide	Decomposes at room temperature, rare, not a significant pollutant.

NO_x is expressed as NO_2. The molecular weight of NO_2 is 46.01 whereas the molecular weight weight of NO is 30.01. Expressing the quantity of NO present as NO_2 requires that the amount of NO be multiplied by the ratio of molecular weight, 46.01/30.01. For instance, if 10 pounds of NO_2 and 150 pounds of NO are present, the quantity of NO_x, expressed as NO_2, is the sum of: NO present times the ratio of weight of NO_2/NO, i.e.,

$$150 \text{ lbs NO} \cdot \frac{46.01 \text{ lb } NO_2}{30.01 \text{ lb NO}} = 299.7 \text{ lb } NO_2 \text{ equivalent}$$

plus NO_2 present, 10.0 lb

Total, NO_x 309.7 lb,expressed as NO_2

It is important to note that knowing only the quantity of NO_x present it is not possible to estimate the NO and/or the NO_2 components. The concept of NO_x is a convenient means of describing the magnitude of the level of nitrogen oxide pollutants in a gas stream.

If the NO and NO_2 components of a gas stream must be determined and the NO_x quantity is known, by assuming a ratio of NO to NO_2, these component values can be established. For example:

A gas stream contains 1000 pounds of NO_x, expressed as NO_2 with a ratio of NO to NO_2 of 100:1, determine the quantities of NO and NO_2 present.

With the NO_2 quantity as y and the NO quantity as $100y$,

$$\text{equivalent to } \frac{46.01}{30.01} \cdot 100y = 153.32y \text{ NO}_2,$$

the total NO_2 equivalent is $154.32y$ lb.
For a total flow of 1000 lb = $154.32y$,
$y = 6.48$ lb NO_2
$100y = 648.00$ lb NO

Sulfur is released into the atmosphere from burning processes in the form of sulfur dioxide, SO_2, and sulfur trioxide, SO_3. More than 95% of the sulfur oxides generated are SO_2 which will slowly oxidize to SO_3. Sulfur trioxide is highly soluble in water and forms sulfuric acid, H_2SO_4, in water. A summary of the effects of both of these compounds is listed in Table 5–2.

As noted above, sulfur oxides are significant pollutants. They will spot and bleach plants and trees and they have been found, in concentrations as low as 0.5 ppm, to cause damage to fruit trees such as apple and pear. Alfalfa, barley, and various species of pine and other conifers are also sensitive to the presence of sulfur oxides.

Rain will wash sulfur oxides from the atmosphere, but the rain will turn acidic. Acid rain is increasingly of concern because of its detrimental effect on plant life, fabrics, metals, and other structural materials. It is a major cause of deterioration of statuary, building facades, and other structures throughout the industrial (and semi-industrial) world.

In human life, sulfur dioxide is an eye irritant, causes and aggravates respiratory diseases such as emphysema and bronchitis, and studies have indicated that there may be a link between sulfur oxides and the occurrence of lung cancer.

Beside specific direct pollutant effects of sulfur oxides, they are contributors to gross atomospheric effects such as smog and haze.

Table 5–2 Sulfur Oxides

Formula	Name	Effects
SO_2	Sulfur Dioxide	First product of combustion. Accounts for 95% of man's emissions of sulfur to the atmosphere. Changes slowly to SO_3 or if dissolved oxidizes slowly to SO_4—and can form sulfurous acid, H_2SO_3.
SO_3	Sulfur Trioxide	Readily dissolves in water to form sulfuric acid, H_2SO_4. Sulfuric acid may be present in the air as an aerosol. It reacts with other pollutants to form sulfate aerosols.

Another sulfur family pollutant is hydrogen sulfide, H_2S. This is an extremely odorous compound, detectable in the amounts of parts per billion in the atmosphere. Hydrogen sulfide and mercaptans, a family of organic sulfur compounds, are detectable by odor long before their concentration can become a danger to health. There is evidence that these compounds are harmful to human health. (In actuality, hydrogen sulfide in lethal concentrations paralyzes the olifactory nerve in seconds preventing detection in time to save the observer.) Hydrogen sulfide acts as an oxygen scavenger, much in the way that carbon monixide does. It seeks to oxidize to a sulfate and water:

$$2\,H_2S + 5\,O_2 \rightarrow 2\,H_2O + 2\,SO_4^{-}$$

The water will normally vaporize and the sulfates are aerosols, small solid particulate matter within the atmosphere. Hydrogen sulfide is not generated by industry in appreciable quantities. Three industries which do generate this gas are petroleum processing, sulfur recovery, and kraft paper production. It is a product of organic or biological processes, associated with the unpleasant odors of marshes, skunks, rotten eggs, and sewer gas.

Organic Gas Discharges

The majority of organic discharges into the atmosphere occur from natural sources and transportation activities. Of the many organic discharges from industrial sources the more significant ones are as follows:

- Oxygenated hydrocarbons which include aldehydes, ketones, alcohols, and acids. In sufficient quantity they will produce eye irritation, reduce visibility in the atmosphere, and will react with other components of the atmosphere to form additional pollutants. Formaldehyde, one of the more common oxygenated hydrocarbons found in industrial discharges, has been known to cause eye irritation in concentrations as low as 0.25 ppm.
- Halogenated hydrocarbons such as carbon tetrachloride, perchloroethylene, etc., may contribute to atmospheric clarity problems and may also generate odor.
- Olefins, a group of unsaturated hydrocarbon compounds, readily react with many other chemical compounds. (Unsaturated compounds are those where the carbon atoms are not exhibiting their maximum valence capability. Adjacent carbon atoms will share double bonds. In saturated compounds carbon atoms in adjacent locations will share only one bond.) Most olefins appear to have no direct effect on animal life in small concentrations. Some of these compounds have been found to cause a general reduction in plant growth. Olefins do take

part in photochemical reactions in the presence of nitrogen oxides and several other pollutants.

- Aromatics, including benzene, toluene, and xylene. Some of these compounds have been found to be carcinogenic. One of the most toxic of these occurring in industrial processes is benzo(a)pyrene (also written as 3,4, benzopyrene). This compound is relatively simple to detect, and because it is highly toxic and is widespread, is used as a standard of emissions whereby the level of hydrocarbons is ascertained. A primary source of these compounds is the incomplete combustion of organic materials, particularly from transportation vehicles.

The most apparent effect of organic gas discharges is their effect on the atmosphere. They react with other elements of the atmosphere to form photochemical smog which is as much a danger to health as it is to the aesthetic quality of an area.

Atmospheric Effects

Atmospheric effects of pollution include the formation of photochemical smog and acid rain. Acid rain, as noted above in the discussion of sulfur discharges, will form in the presence of sulfur oxides. Sulfur dioxide will oxidize to sulfur trioxide and this element is highly soluble in water. It will form sulfuric acid upon contact with water; that is, falling rain. The sulfur oxides may also be carried by prevailing winds to distant areas before carried by rains to earth. It also has the potential for nucleating upon rising water vapor particles forming clouds with relatively high levels of potential acid rain. Again, these clouds may travel long distances before their acid rain discharges to the ground.

Photochemical smog formation is an extremely complex process. It's formation is not completely understood but it is known that nitrogen oxides, organic discharges, and water vapor are major participants in its occurrence and intensity. A simplified set of reactions which generate this atmospheric phenomenon is as follows:

$$NO_2 + \text{Sunlight} \rightarrow NO + O \quad (5\text{-}1)$$
$$O + O_2 \rightarrow O_3 \quad (5\text{-}2)$$
$$O_3 + NO \rightarrow NO_2 + O_2 \quad (5\text{-}3)$$
$$HO + RH(+O_2) \rightarrow RO_2 \quad (5\text{-}4)$$
$$RO_2 + NO \rightarrow RO + NO_2 \quad (5\text{-}5)$$
$$RO + O_2 \rightarrow HO_2 + \text{aldehydes, etc.} \quad (5\text{-}6)$$
$$HO_2 + NO \rightarrow HO + NO_2 \quad (5\text{-}7)$$
$$O + SO_2 \rightarrow SO_3 \quad (5\text{-}8)$$
$$SO_3 + H_2O \rightarrow H_2SO_4 \quad (5\text{-}9)$$

The significant components of smog, those components that cause haze and lachrymatory (tear-inducing) effects are nitrogen dioxide (NO_2) and sulfuric acid (H_2SO_4). Ozone (O_3), is formed by the action of sunlight on nitrogen dioxide, equation 5-1, and from ozone, the free oxygen radical, O becomes available. It has been found that the intensity of a smog episode is directly related to the amount of ozone present in the atmosphere. In Los Angeles, a serious occurrence of smog is characterized by a level of ozone of at least 0.5 ppm by volume (1070μg/cm by weight).

Examining the above equations further, the radical HO is formed when water is in the presence of certain hydrocarbons. The symbol "R" represents a hydrocarbon radical, one generated from an organic emission such as an aldehyde or aromatic derivative. The reactions in equations 5-3, 5-5 and 5-7 produce nitrogen dioxide. Equations 5-1, 5-2, 5-4 and 5-6 describe some of the reactions which produce the free oxygen, ozone, or other elements that promote the formation of nitrogen dioxide. For instance, a series of aldehydes (plus ketones and other hydrocarbons) is generated from RO, equation 5-6, where these hydrocarbons then become available for generating nitrogen dioxide from nitrogen oxide.

Normally nitrogen oxide is generated from combustion sources in far greater quantities than nitrogen dioxide and its natural conversion to nitrogen dioxide is relatively slow. With the above processes this conversion is significantly more rapid and is more widespread. (The generation of NO and/or NO_2 is negligible at temperatures below 2000°F. The majority of incineration processes operate at temperatures below this level and, therefore, produce little NO_x.)

Equations 5-8 and 5-9 describe the generation of sulfuric acid in the atmosphere. This is a relatively straightforward process resulting from the high solubility of sulfur trioxide in water. Over 95% of the sulfur oxide emissions from a combustion process are in the form of sulfur dioxide and the presence of ozone accelerates the formation of the soluble sulfur trioxide from the sulfur dioxide present.

References and Bibliography

1. Skizim, D. Gaseous emission control is vital. *Solid Wastes Management* (April 1982).
2. Innes, W. Effect of nitrogen oxide emissions on ozone levels in metropolitan regions. *Environmental Science and Technology* (August 1981).
3. Bagwell, F. Oxides of nitrogen emission reduction program for oil and gas fired utility boilers. *Engineering Science and Technology* (April 1980).
4. Rollins, R., and J. Homolya. Measurement of gaseous hydrogen chloride emissions from municipal refuse energy recovery systems in the United States. *Environmental Science and Technology*, (November 1979).
5. Jahnke, J. A research study of gaseous emissions from a municipal incinerator. *APCA Journal* (August 1977).

6

Dioxins

The term *dioxin* is a generalization of a family of chlorinated organic compounds, some of which have been found to be extremely hazardous to life in minute quantities. There has been public concern that these compounds may be generated from incineration processes.

Dioxin Family

The dioxin molecular framework consists of two benzene rings connected by two oxygen bridges, as shown in Figure 6–1. Also indicated in this illustration are the basic molecular frameworks of the related compounds polychlorinated dibenzofurans (PDFFS) and polychlorinated biphenols (PCBs).

At least two chlorine atoms occurring at two or more of the eight numbered locations define a dioxin, and there are 73 different combinations, as noted in Table 6–1. These are all known as polychlorinated dibenzo-p-dioxins, (PCDDs). There are two dioxin isomers possible with

Table 6–1 Dioxin Analyses

Polychlorinated Dibenzo-p-dioxin	Molecular Formula	Average Molecular Weight	No. of Isomers
Dichloro-dibenzo-p-dioxin (DCDD)	$C_{12}H_6Cl_2O_2$	253.1	10
Trichloro-dibenzo-p-dioxin (Tri-CDD)	$C_{12}H_5Cl_3O_2$	287.5	14
Tetrachloro-dibenzo-p-dioxin (TCDD)	$C_{12}H_4Cl_4O_2$	322.0	22
Pentachloro-dibenzo-p-dioxin (Penta-CDD)	$C_{12}H_3Cl_5O_2$	356.4	14
Hexachloro-dibenzo-p-dioxin (Hexa-CDD)	$C_{12}H_2Cl_6O_2$	390.9	10
Heptachloro-dibenzo-p-dioxin (Hepta-CDD)	$C_{12}HCl_7O_2$	425.3	2
Octachloro-dibenzo-p-dioxin (OCDD)	$C_{12}Cl_8O_2$	459.8	1
		Total Isomers	73

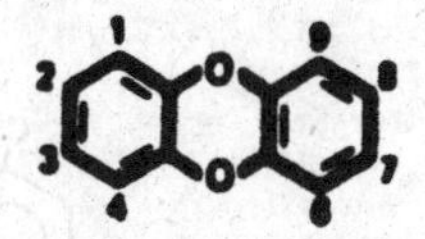

DIBENZO-P-DIOXIN

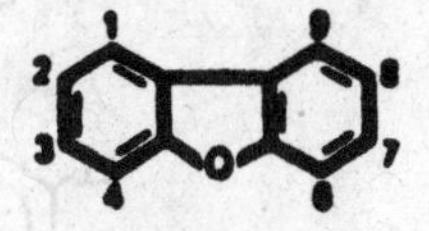

DIBENZOFURAN

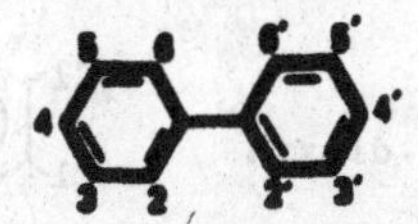

POLYCHLORINATED BYPHÉNYL (With the addition of at least two atoms of chlorine)

Figure 6–1 Polycyclic organic compounds.

only one chlorine atom, monochlorinated dibenzo-p-dioxin. The general term, which includes the mono- and poly- chlorinated dioxins, is *chlorinated dibenzo-p-dioxin*, (CDD).

Of these varieties of dioxins, the most toxic dioxin found to date is 2,3,7,8-TCDD, which has the molecular structure illustrated in Figure 6–2 and an average molecular weight of approximately 322. This compound is a solid at standard conditions with a melting point in the range of 577°F to 581°F. At 77°F its solubility in water is 0.0002 mg/l, 57 mg/l in benzene, 4.8 mg/l in octanol and 1.0 mg/l in methanol.

PCDDs are thermally stable up through 1300°F. Above this temperature they will start to decompose. Their vapor pressure is extremely low, less than one millionth of a millimeter of mercury at standard conditions.

This low vapor pressure indicates that a PCDD compound will not readily volatilize from a surface at ambient temperatures. This physical

PCDD

Dichloro-dibenzo-p-dioxin

Trichloro-dibenzo-p-dioxin

Tetrachloro-dibenzo-p-dioxin

Pentachloro-dibenzo-p-dioxin

Hexachloro-dibenzo-p-dioxin

Heptachloro-dibenzo-p-dioxin

Octachloro-dibenzo-p-dioxin

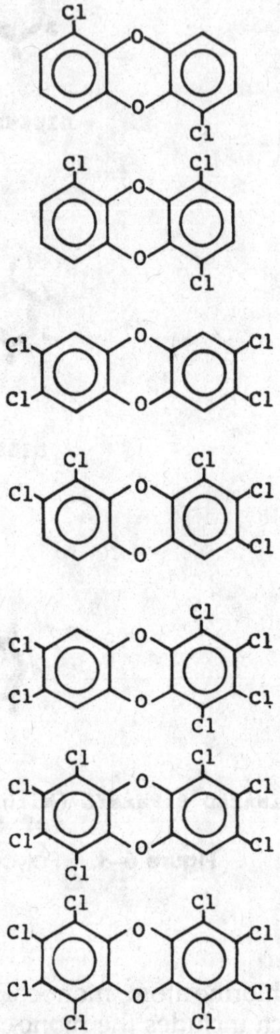

Figure 6–2 The dioxin family.

property, plus thermal stability and low solubility in water are three of the more important qualities of dioxins with respect to their fate in the environment.

Polychlorinated dibenzofurans have a chemistry and structure similar to that of dioxins. Figure 6–2 illustrates the PCDF molecule. The higher toxicity of PCDD and the apparently greater incidence of it from manufacturing and other processes than PCDF, has focused greater attention on it. PCDF, however, should be considered to be of the same magnitude

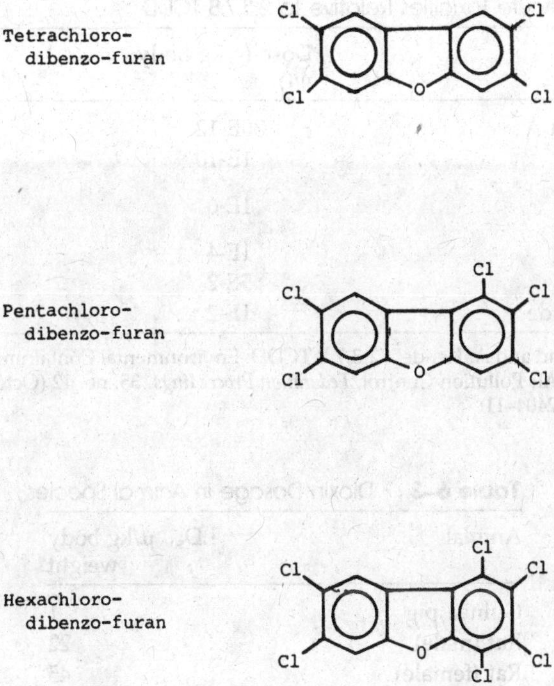

Tetrachloro-
dibenzo-furan

Cl Cl
O
Cl Cl

Pentachloro-
dibenzo-furan

Cl Cl Cl
O
Cl Cl

Hexachloro-
dibenzo-furan

Cl Cl Cl
O
Cl Cl Cl

Figure 6–3 Typical Furan Molecules

of danger as PCDD and the following discussions of PCDD should be applied equally to PCDF.

Toxicity

Although no cases of human death or even long-term disability have been attributed to dioxins in the United States or elsewhere, there is widespread fear among the public at large that dioxins pose a severe health threat to human life. In the 1970s the compound 2,3,7,8 TCDD was isolated and was found to be extremely toxic to small animals, as illustrated in Table 6–2. It is the most toxic synthetic chemical known, 500 times more potent than strychnine and 10,000 times more potent than cyanide as determined by laboratory analyses.

The lethal animal dose of PCDDs varies from one animal species to another, as indicated in Table 6–3. The guinea pig and rat, with roughly the same body weight as the hamster, are more sensitive to PCDD toxicity by many orders of magnitude than is the hamster. Similar inconsistencies appear with respect to the dog and the monkey. These differences in animal toxicity make extrapolation to human toxicity a difficult, if not impossible, task.

Table 6–2 Acute Toxicities Relative to 2,3,7,8 TCDD

	Dose (g/kg body wt)	Relative Toxicity
Botulism toxin A	30E-12	35,000
Tetanus toxin	1E-10	10,000
2,3,7,8 TCDD	1E-6	1
1,2,3,4 TCDD	1E-4	.01
Strychnine	5E-2	.002
Sodium Cyanide	1E-2	.0001

Source: A. Poland and A. Kende, "2,3,7,8 TCDD: Environmental Containment and Molecular Probe," In Air Pollution Control, *Federation Proceedings*, 35, no. 12 (October 1976), San Francisco, CA. 2404–11.

Table 6–3 Dioxin Dosage in Animal Species

Animal	LD_{50}, μ/kg body weight
Guinea pig	1
Rat (male)	22
Rat (female)	45
Monkey	<70
Rabbit	115
Mouse	114
Dog	>300
Bullfrog	>500
Hamster	5,000

Source: A. Poland and R. Knudsen, *Annual Review of Pharmacology and Toxicology* (1982), Unpublished Report.

Laboratory animals exhibit a number of toxic effects from PCDDs as follows:

- In the rat and the mouse, 2,3,7,8 TCDD appear to be teratogenic and carcinogenic.
- In the rat, mouse, and monkey (Rhesus) 2,3,7,8 TCDD affects the reproductive system by inducing abortions, reducing the ability to conceive, and causing fetotoxicity.
- In mice, 2,3,7,8 TCDD causes immunotoxicity.

Many of the procedures utilized to obtain data on animal toxicity resulted in a wide scatter. The results varied over a relatively wide range. One reason for this is that dioxins are being measured in extremely small quantities, parts per billion or parts per trillion. Minor changes in methods or procedures from one laboratory to another could result in major differences in the resultant measurement, which is reflected in the above data.

Human Health Effects

Research into the health effects of dioxins on human life will be continuing for many years. Dioxins are by-products in the manufacture of agricultural and industrial chemicals. The only known human contact has been accidental. No experiments have been performed, and none can be performed, to determine human toxicity. Purposeful human exposure to dioxins, based on its known toxicity in laboratory animals, is a moral anathema. The incidents of accidental exposure do not lend themselves to rigorous analysis. In many cases the prior medical history of the people exposed to dioxins has never been established, which is a vital factor in evaluating the effect of an abnormal occurrence.

Table 6–4 lists some of the known occurrences of possible human exposure to dioxins. There are a number of conclusions that have been drawn from these episodes concerning dioxin effects on humans:

- In each case of exposure a skin disorder, chloracne, has occurred. This is a severe facial eruption, unsightly, and is a severe irritant.
- Liver damage has been reported a number of months after exposure; however, in every reported case, signs of this damage have disappeared within a period of 2 to 5 years.
- No human symptoms or reactions have been found to be permanently injurious to well-being.
- Although no relationship to malignancies has been found between dioxins and human life, this is a fear, particularly as a long-term effect.

Generation of Dioxins

Until the threshold limit of dioxin toxicity is determined, the mechanism of dioxin generation, in any measureable amount, must be investigated and understood. There are many theories of dioxin formation from incineration including the following:

- Burning of wastes which contain trace levels of PCDD will necessarily produce PCDD in the exhaust stream.
- The presence of two or more chlorinated organics act as precursors in the formation of PCDD. By a process termed *dimerization* these

Table 6–4 Human Exposure to Dioxins, Notable Incidents

Year	Location	Deaths: Actual/Expected
1949–77	Sweden, railroad workers	18/25
1949–76	Finland, spraying insecticide	normal
1949	West Virginia, Nitro, industrial accident	32/46
1950–71	Midland, Michigan, 2,4,5,T production workers	11/20
1952–53	Boehringer, Germany	–
1953	Ludwigshaven, Germany industrial accident	17/1–25
1954	Boehringer, Hamburg, Germany	–
1956	Grenoble, France, industrial accident	–
1956	New Jersey, 2,4,5,T production workers	–
1962–71	Vietnam, agent orange exposure	–
1963	Amsterdam, Holland, industrial accident	3/–
1964	USSR, production workers	–
1964	Midland, Michigan, industrial accident	4/7–8
1965–68	Czechoslavakia, 2,4,5,T production workers	5/
1966	Grenoble, France, industrial accident	–
1968	Bolsover, England, industrial accident	–
1970	England, laboratory workers making PCDD's	–
1972–73	Austria, production workers	–
1974	Germany, industrial accident	–
1976	Seveso, Italy, industrial accident	normal
1978	Alsea, Oregon, local female residents	–

Source: C. Kemp. "Notes on Polychloro Dibenzodioxins and Polychloro Dibenzofurans in Connection With Waste-to-Energy Plants," Browning Ferris Industries, April 1983.

compounds (such as chlorinated phenols) will combine, under appropriate conditions of temperature and oxygen availability, to form PCDD.

- PCDD may be formed by partial oxidation of single-molecule precursor compounds, such as the partial oxidation of PCBs (polychlorinated biphenols).
- The presence of chlorine and the chlorine (chloride) attack of basic hydrocarbon (aromatic) structures associated with lignin, such as wood, vegetable residues, etc., encourage PCDD formation. For instance, the attack of partially pyrolyzed char particles by HCl or other chlorine sources present in the flue gas stream due to the incineration of chlorinated plastics, or reactions from inorganic chloride salts, may produce PCDD emissions. In particular, PCDD/PCDF compounds are found when burning wood, which has a significant lignin component, whereas essentially no such emissions have been found when burning paper, which is lignin-free for the most part.

Estimating Dioxin Emissions

Because of the highly toxic nature of PCDDs and PCDFs, concern has developed over the generation of minute quantities of these substances and their release into the environment. In an effort to determine their presence down to parts per billion, or even less, questions have arisen with regard to the accuracy and repeatability of the analytical techniques used. The available data, therefore, should not be used as absolute values. In addition, the method of collection of this data and the equipment operating parameters which result in PCDD and PCDF generation have, in general, not been fully documented.

Table 6–5 lists dioxins found in various areas of the environment. It appears to be fairly ubiquitous, detected in the home, in the earth, in the air, and in the discharge from many different types of combustion sources.

The values in Table 6–6 indicate that PCDDs and PCDFs are, indeed, produced from the burning of refuse. These figures are presented in micrograms of CDD per metric ton of refuse charged, which is a ratio of 10^{12} to 1. The ratio of 100 micrograms per ton is equivalent to less than one person from the entire population (4 billion persons) of this planet.

A number of theories have been proposed to try to explain the apparent difference between data from European and American sources (see Table 6–6), which indicate higher dioxin levels in Europe. There is concern that the sample size is not sufficiently large to justify a statistical conclusion. Another explanation is that in most cases a control was not established to provide true, unbiased results from the stack gas sampled. Analytical methods between the United States and Europe differ, and with the extremely small quantities of organics measured, differing results could be possible. Assuming that the analytical and sampling techniques

Table 6–5 Chlorinated Dioxins in the Enrivonment

Sample	Apparent Dioxin Content, ppb			
	TCDD	HCDD	H_7CDD	OCDD
Soil				
Rural	*	*	* – .05	* – .2
Urban (Lansing, MI)	*	.03–1.2	.03–2	.05–2
Urban (Chicago)	.005–.03	.03–.3	.1–3	.4–22
Dow Chemical (Michigan)	1–120	7–280	70–3,200	490–20,000
Dust				
Dow Chemical Laboratory	1–4	9–35	140–1,200	650–7,500
Midland, MI	.03–.04	.2–.4	2–4	20–30
Detroit, MI	* – .03	* – .3	.3–4	.1–4
St. Louis, MO	.3	2	34	210
Chicago, IL	.04	* – .3	.6–3	3–8
Wastewater Treatment Sludge				
Milorganite (Milwaukee)	.31	2	30	180
Incinerators				
Dow Powerhouse	38	2	4	24
Dow Rotary Incin. Stack	*	1–5	4–100	9–950
Dow Tar Burner	*	1–20	27–160	190–440
Nashville Incinerator	7.7	14	28	30
European Incinerators	2–20	30–200	60–130	40–120
Mufflers				
Diesel Truck Muffler	.023	.020	.100	.26
Auto Muffler	* – .008	*	.003–.01	.02–.07
Other sources				
Home Fireplace Soot	* – .4	.2–3	.7–16	.9–25
Home electrostatic Prec.	*	.004–.008	.009	.02–.05
Charcoal Broiled Steak	*	*	*	.03

*Not detected

Source: R. Bumb, et al., "Trace Chemistries of Fire," *Science* 210: (October 24, 1980).

were satisfactory, and compatible, the differences may be caused by the differences in the wood to paper ratio in European and American refuse. As discussed previously, it is believed that lignin contributes to the generation of PCDD and PCDF in a furnace atmosphere. Europe has, in general, a higher proportion of wood to paper in their refuse than

Table 6–6 PCDD and PCDF Emissions From Incinerators in the U.S.A. and Europe (μg/tonne MSW)

Compound	(a)	(b)	(c)	(d)	(e)	Mean	U.S.A.
TrCDD	*	–	*	*	*	*	65
TCDD	610	–	250	175	26	266	32
PnCDD	2,418	–	467	452	72	853	–
HxCDD	3,263	–	642	766	164	1,209	82
HpCDD	3,366	–	581	766	158	1,218	37
OCDD	1,899	–	704	175	323	776	13
Total	11,556	2,563	2,644	2,333	743	4,322	229
TrCDF	*	–	*	*	*	*	1,529
TCDF	1,562	–	700	260	146	667	446
PnCDF	2,704	–	743	541	179	1,042	–
HxCDF	4,786	–	804	906	123	1,655	306
HpCDF	2,614	–	880	640	82	1,054	37
OCDF	581	–	590	80	54	326	3
Total	12,247	4,159	3,717	2,427	584	4,744	2,321

Note: *Not detected

The parenthetical values (a) through (e) refer to tests performed on five different European incinerators.

Source: S. Graham et al., *Production of Polychlorinated Dibenzo-dioxins (PCDD) and Furans (PCDF) From Resource Recovery Facilities*, Part I, The American Society of Mechanical Engineers, Solid Waste Processing Division, *Proceedings*, 1984.

America and this means there is a higher lignin content entering their incinerators.

Dioxin Regulations

The regulation of dioxins under the Resource Conservation and Recovery Act (RCRA) was proposed in the April 4, 1983 issue of the *Federal Register*. There are a number of important provisos in this proposed regulation, including the following:

- All chlorinated dibenzo-p-dioxins (CDDs) will be considered equally hazardous. Both of these families of compounds will be considered as "acute hazardous wastes" and will be included in the Appendix 8 (hazardous constituents) hazardous waste listing (see chapter 2).
- CDD and CDF compounds can be incinerated only at fully authorized facilities which have permits demonstrating the capability for 99.99%

destruction and removal (DRE) efficiency for principal organic hazardous constituents (POHCs) which are at least as difficult to incinerate as CDDs or CDFs. Such other compounds would include carbon tetrachloride and pentachlorophenol.

In addition to federal concerns, there are local concerns in the states where dioxin contamination has been found and has been widely publicized, such as Missouri and Michigan. These states, and others, can be expected to develop their own statutes which will be more severe than federal legislation and which will be tailored specifically to local conditions and local remediation.

References and Bibliography

1. Shih, C., Ackerman, D., Scinto, L., et al, 1980 *POM emissions from stationary conventional combustion sources, with emphasis on polychlorinated compounds of dibenzo-p-dioxin (PCDD's), biphenyl (PCB's) and dibenzofuran (PCDF's)*, preliminary draft, TRW, Inc. EPA Contract No. 68-02-3138, January.
2. Arthur D. Little, Inc. 1981. *Dioxin from combustion sources*. New York: The American Society of Mechanical Engineers.
3. Shaub, W., and W. Tsang. Dioxin formation in incinerators. *Environmental Science and Technology* (1983) 17:12.
4. Kemp, C. April 1983. Notes on polychloro dibenzodioxins polychloro dibenzofurans in connection with waste-to-energy plants. Browning Ferris Industries. Unpublished.
5. Tiernan, T. and F. Dryden. 1980. *Dioxins*. USEPA-600/2-80-197, Washington, DC: Government Printing Office, November.
6. Shaub, W. 1984. *Containment of dioxin emissions from refuse fired thermal processing units—Prospects and technical issues*. US Department of Commerce, NBSIR 84-2872. Washington, DC: Government Printing Office, May.
7. Thompson, D., and M. Cullinane. 1983. *Evaluation of onsite incineration as a remedial action alternative for dioxin contaminated sites*. Working draft, U. S. Army Engineer Waterways Experiment Station, USEPA IAG No. DW930176-01-0. Washington, DC; Government Printing Office, August.
8. Young, A., et al. 1978. The toxicology, environmental fate and human risk of herbicide orange and its associated dioxin, U.S. Air Force occupational and environmental health laboratory. *USAF Report OEHL TR-78-92*. Washington, DC: Government Printing Office, October.
9. Buser, H. Formation of polychlorinated dibenzofurans (PCDF's) from the pyrolysis of individual PCB isomers. *Chemosphere*, (1979) 3 157–174.
10. Junk, G. and J. Richard. Dioxins not detected in effluents from coal/refuse combustion. *Chemosphere*. (1981) 10. 11–12: 1237–1241.

11. Rappe, C., S. Marklund, et al. Formation of polychlorinated dibenzo-p-dioxins (PCDD's) and dibenzofurans (PCDF's) by burning or heating chlorophenates. *Chemosphere*. (1978) 3: 269–281.
12. Bumb, R.,et al. Trace chemistries of fire: A source of chlorinated dioxins. *Science* (October 1980) Vol. 210.
13. Duckett, E. Dioxins in perspective: Knowns, unknowns, resolving the issues. *Solid Wastes Management*. (May 1981).
14. Dioxin Report. *Chemical and Engineering News* (June 6, 1983).
15. Cavallaro, A., L. Luciani, G., Ceroni, et al. Summary of Results of $PCDD_5$ Analyses From Incinerator Effluents, *Chemisphere*. (1982) 11/9: 859.
16.Chlorinated dioxins and furans in the environment, *Environmental Science and Technology*. (1983) 17/3:124A.
17. Young, A., H. Kang, and B. Shepard. Chlorinated dioxins as herbicide contaminants. *Environmental Science and Technology* (1983) 17/11:530A–540A.
18. Main, J. Dow vs. the dioxin monster, *Fortune* (May 30,1983).
19. Eiceman, M., R. Clement, and F. Karasek. Analysis of fly ash from municipal incinerators for Trace Organic Compounds, *Analytical Chemistry*, (December 1979) 51:14.
20. Gizzi, F., et al. Polychlorinated dibenzo-p-dioxins (PCDD) and polychlorinated dibenzofurane (PCDF) in Emissions From An Urban Incinerator (Average and Peak Values), *Chemosphere* (1982) 11/6:557–583.
21. Chlorodibenzo-p-dioxins and chlorodibenzofurans are trace components of fly ash and flue gas of some municipal incinerators in the Netherlands. Unpublished article, undated.
22. Lustenhouwer, J., K. Olie, and O. Hutzinger. Chlorinated dibenzo-p-dioxins and related compounds in incinerator effluents: A review of measurements and mechanisms of formation. *Chemosphere* (1980) 9: 501–522.
23. Graham, S., et al. Production of polychlorinated dibenzo-p-dioxins (PCDD) and furans (PCDF) from resource recovery facilities, Part I, The American Society of Mechanical Engineers, *Solid Waste Processing Division, Proceedings*, Orlando, FL, 1984.
24. Niessen, W. Production of polychlorinated dibenzo-p-dioxins (PCDD) and furan (PCDF) from resource recovery facilities, Part II, The American Society of Mechanical Engineers, *Solid Waste Processing Division, Proceedings*, Orlando, Fl. 1984.

7

Odor Emissions

Odors associated with incineration are normally organic and result from the incomplete combustion of organic matter in the waste feed. The most effective means of dealing with the problem of odor generation is alteration of the burning process to increase burning efficiency, thereby decreasing the occurrence of odor-causing compounds in the exhaust stream.

Unfortunately, waste streams are not consistent in chemistry or quality. As waste quality changes, the efficiency of the waste burning process for that particular waste will change, and combustion parameters will have to be adjusted to reduce the possibility of odor generation. Changes in the burning process may not be practical with a waste that changes in character within a short period of time. External mechanisms of odor control may be required. An understanding of the nature of odor and odor quantification is necessary prior to a discussion of methods of odor control.

Characteristics of Odor

Odor is defined as that sensation perceived by the human nose. An odorant is a substance that stimulates an olfactory response, the sensation of smell. It is the substance that causes an odor. The odorant is a gas or a vapor but it can originate from a solid, a liquid, or a concentrated gas.

Methods developed to attempt to describe an odor in discrete terms make use of the four characteristics of an odor, as follows:

Intensity	The strength of an odor.
Pervasiveness	How much air the odor can pervade or be diluted with and still be detectable.
Acceptability	Whether the smell is perceived as pleasant or unpleasant
Quality	What the odor smells like; for example, flowers, exhaust, baked goods, etc.

Intensity and pervasiveness are measurable, making use of an "odor panel" and establishing "odor units" as described in the next section of this chapter. There are many odor quality and acceptability classifications and a number of these will be described below.

Measurement of Odor Intensity

The only reliable means for the detection of odor is the human nose. In an attempt to define and classify odor, utilizing the human olifactory response, an odor measurement procedure has been established (ASTM Specification D-1391) that has wide acceptance throughout the country. This procedure involves the following steps:

- A group of people (6 to 12) is chosen to serve on an "odor panel."
- Odor samples are brought to the odor panel, remote from the source of odor, and they decide whether or not they can perceive any odor from these samples.
- Pleasantness or unpleasantness of an odor is not a judgment factor, just intensity.
- Odor units are established. One odor unit is defined as the amount of odor in 1 cubic foot of air that half a panel of odor judges can smell and half cannot. The number of dilutions necessary to obtain one odor unit is equal to the number of odor units for a particular sample. One odor unit also defines the "threshold" of odor detection.

Threshold Values

The concentration of a contaminant in air that can be detected as an odor is extremely small. Less than three parts of contiminant per billion parts of air can be detected with some compounds such as amines and acrylates. Typical odor threshold values are listed in Table 7–1. The first column of values is the threshold level of half of an odor panel. The second column is that level of odor, or threshold, that can be detected by 100% of the odor panel.

With these extremely low levels of odor threshold values, an odor control system for these listed contaminants must remove essentially all of the odor-causing compound to eliminate the presence of the odor.

Odor Quality

Odor quality and acceptability are extremely subjective criteria. While in most cases the pleasantness or unpleasantness of an odor is agreed upon by a group of observers, the quality of that odor is subject to individual intepretations. The following odor classification systems are the more common of those that have been developed:

Table 7–1 Typical Odor Threshold Values

Compound (Increasing Threshold)	Threshold Value, PPM (Volume) 50% Response	100% Response
Amine, Trimethyl	0.00021	0.00021
Ethyl Acrylate	0.0001	0.00047
Hydrogen Sulfide Gas	0.00021	0.00047
Butyric Acid	0.00047	0.001
Ethyl Mercaptan	0.00047	0.001
p-Cresol	0.00047	0.001
Dimethyl Sulfide	0.001	0.001
Sulfur Dichloride	0.001	0.001
Benzyl Sulfide	0.0021	0.0021
Methyl Mercaptan	0.001	0.0021
Diphenyl Sulfide	0.0021	0.0047
Nitrobenzene	0.0047	0.0047
Pyridine	0.01	0.021
Amine, Monomethyl	0.021	0.021
Phosphine	0.021	0.021
Amine, Dimethyl	0.021	0.047
Benzyl Chloride	0.01	0.047
Bromine	0.047	0.047
Chloral	0.047	0.047
Phenol	0.021	0.047
Diphenyl Ether	0.1	0.1
Styrene	0.047	0.1
Acetaldehyde	0.21	0.21
Acrolein	0.1	0.21
Carbon Disulfide	0.1	0.21
Methyl Methacrylate	0.21	0.21
Monochlorobenzene	0.21	0.21
Chlorine	0.314	0.314
Allyl Chloride	0.21	0.47
Methyl Isobutyl Ketone	0.47	0.47
p-Xylene	0.47	0.47
Sulfur Dioxide	0.47	0.47
Acetic Acid	0.21	1.0
Aniline	1.0	1.0
Phosgene	0.47	1.0
Formaldehyde	1.0	1.0
Toluene Diisocyanate	0.21	2.14
Toluene (from petroleum)	2.14	2.14
Benzene	2.14	4.68
Toluene (from coke)	2.14	4.68
Perchloroethylene	4.68	4.68
Ethanol (synthetic)	4.68	10.0
Methyl Ethyl Ketone	4.68	10.0

Table 7–1 Typical Odor Threshold Values (*Continued*)

Compound (Increasing Threshold)	Threshold Value, PPM (Volume) 50% Response	100% Response
Hydrochloric Acid Gas	10.0	10.0
Carbon Tetrachloride	10.0	21.4
Acrylonitrate	21.4	21.4
Trichloroethylene	21.4	21.4
Ammonia	21.4	46.8
Dimethylacetamide	21.4	46.8
Dimethylformamide	21.4	100.0
Acetone	46.8	100.0
Carbon Tetrachloride	46.8	100.0
Methanol	100.0	100.0
Methylene Chloride	214.0	214.0

Source: Manufacturing Chemist's Association, *Research on Chemical Odors* (New York: MCA, October 1978).

Zwaardemaker's Odor Quality Classification System[1]

1. Ethereal or fruity
2. Aromatic
   a. Camphoraceous: borneol, camphor, eucalyptole
   b. Spicy: eugenol, ginger, pepper, cinnamon, cassia, mace
   c. Anise-Lavender: lavender, menthol, thymol, safrole, peppermint, anethole
   d. Lemon-Rose: geraniol, citral, linalyl, acetate, sandalwood
   e. Amygdalin: benzaldehyde, oil of bitter almond, nitrobenzene, prussic acid, salicylaldehyde
3. Fragrant or balsamic
   a. Floral: jasmine, ilang-ilang, orange blossom, lilac, terpineol, lilly of the valley
   b. Lily: tuberose, narcissus, hyacinth, orris, violet, ionone, mignonette
   c. Balsamic: vanillin, piperonal, coumarin, balsams of Peru and Tolu
4. Ambrosial: musk and amber
5. Alliaceous or Garlic:
   a. Alliaceous: hydrides of sulfur, selenium and tellurium and arsenic
   b. Cacodyl fish odors: hydrides of phosphorus and arsenic, cacodyl compounds, trimethylamine
   c. Bromine odors: bromine, chlorine, quinone
6. Empyreumatic or Burned: as in tar, baked bread, roasted coffee tobacco, benzene, napthalene, phenol, and products of the distillation of wood
7. Hircine or Boaty: due to caproic and caprylic esters contained in sweat and also typified by persperation and cheese
8. Repulsive: such as given off by many narcotic plants and by acanthus

9. Nauseating or Fetid: such as given off by products of putrefaction and by certain plants

Another system of odor classification has been developed by S. Horstman,[2] assigns a number in an attempt to quantify odor quality, as follows:

Horstman's Odor Quality Classification

0	Flowers
1	Pulpwood
2	Smoke, Woodsmoke
3	Burning Leaves
4	Mustiness
5	Gasoline
6	Rendering Plant
7	Rubbish Burning
8	Animal Odors
9	Miscellaneous Odors
None	No Odor

A third system that has found acceptance in the classification of odors is as follows:

Henning's Odor Quality Classification[3]

1. Spicy: cloves, cinnamon, nutmeg, etc.
2. Flowery: heliotrope, jasmine, etc.
3. Fruity: apple, orange oil, vinegar, etc.
4. Resinous: coniferous oils and turpentine
5. Foul: hydrogen sulfide and products of decay
6. Burned: tarry and scorched substances

Odor classification provides a convenient means of communicating about odors and is invaluable in masking technology, as will be discussed below.

Odor Control

Techniques for odor control from incineration processes are normally one of the following:

- Fume Incineration (afterburner)
- Packed Tower (absorber)

There are other techniques that have been used for incinerator odor control with varied records of success which include the following:

- Catalytic oxidation
- Adsorption
- Dilution
- Masking

These techniques are discussed below.

Fume Incineration

Burning is the ultimate odor control technique. The majority of odor forming compounds are organic in composition. When oxidized, or burned, they will form carbon dioxide, water vapor, and other innocuous compounds, destroying the odor in the process. For most organic compounds a temperature of 1400°F maintained for a period of at least 0.5 seconds will be satisfactory for complete destruction. Complex organic molecules, such as phenols, require higher temperatures and retention times for effective destruction. In all cases, thermal destruction of any compound also requires effective mixing, or turbulence within the furnace to assure that the temperature/combustion requirements are met throughout the waste.

Separate fume incinerators are often utilized for odor destruction when other techniques are not effective. Destruction of odor by incineration requires the use of supplemental fuel and with the high cost of energy this technique is avoided whenever possible. When it is used, fume incinerators often have energy-saving design features such as recuperative heat exchangers or waste heat boilers. These devices reclaim some of the heating value of the fired fuel, making the incineration option more attractive as a means of odor control.

Catalytic Oxidation

Catalytic oxidation is a relatively low temperature incineration technique. Insertion of a catalytic agent in the gas stream induces destruction of the contaminant at lower temperatures than direct-flame fume incineration. The catalyst normally used is a noble metal compound such as platinum or rhodium. It is used in small quantities, deposited on a support material such as alumina. The noble metal is not used by itself.

This technique is not applicable to all odor compounds, nor to gases with relatively high particulate loadings. Where it can be used, however, it is very effective in odor destruction while providing significant cost reductions over fume incineration.

Chapter 8 "Incineration" provides additional information on catalytic as well as direct-flame incineration of gases.

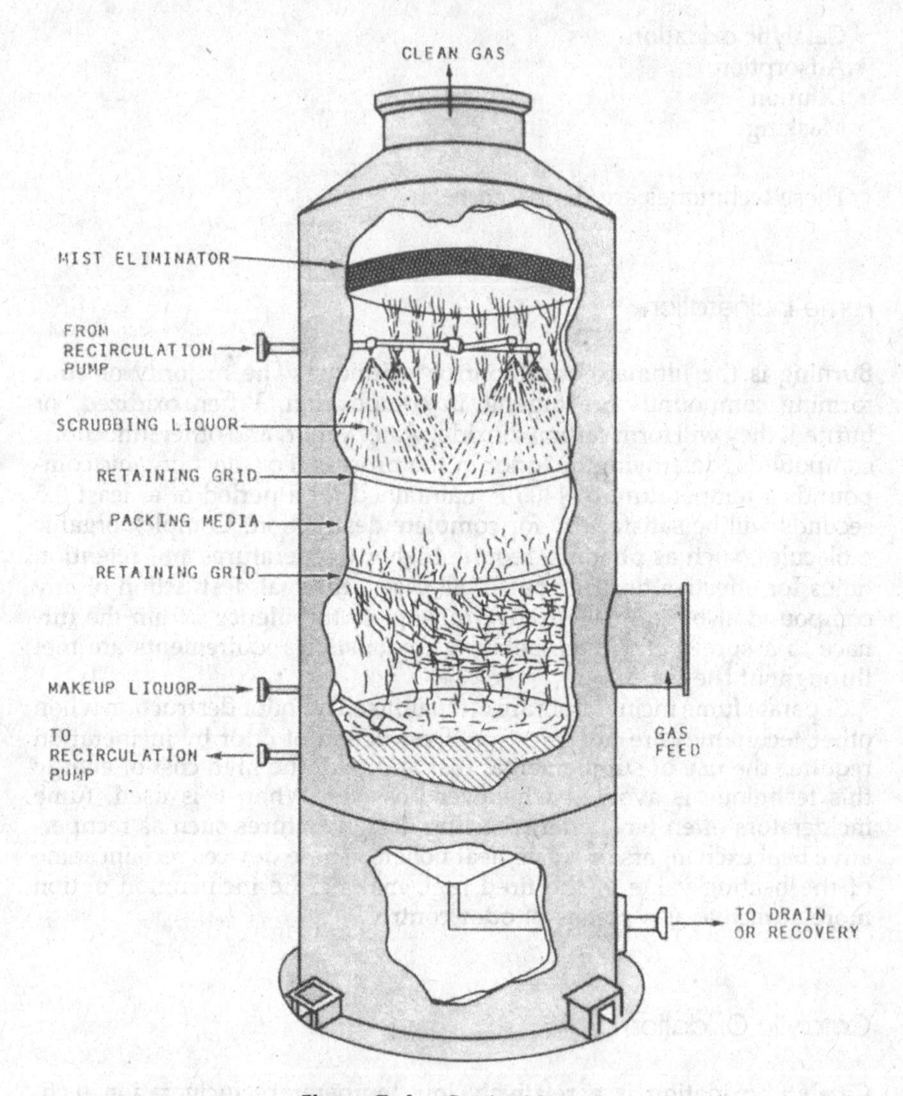

Figure 7–1 Packed tower.

Absorbtion

In this method of odor control the odorous constituent is absorbed into a solution by either chemical bond or solubility. In the case of solubility an odorous gas may be absorbed by a liquid, however, if the odorous component of the gas does not condense and dissolve, the odorous gas may eventually be released. Chapter 4, "Particulate Pollutants" describes sorbtion in greater detail, and these processes are applicable to the gases and aerosols responsible for the presence of odor.

Chemical absorbtion is generally an oxidative process. Many odors are unburned (unoxidized) hydrocarbons. By mixing, or scrubbing with chemicals with the ability to release oxygen (such as potassium permanganate, sodium hypochlorite, or chlorine dioxide) the odorous constituent can be effectively oxidized and likewise lose its odor.

Packed towers such as that shown in Figure 7–1 are commonly used for the absorbtion process. Odorous gas enters the bottom of the tower and rises through a bed of packing, such as that shown in Figure 7–2. Liquid is injected at the top of the tower and the falling liquid wets the packing, providing effective contact and mixing between the odorous gas and the scrubbing liquid. The liquid, which contains the chemical oxidizer, can be collected at the base of the tower and recirculated back through the system.

These systems are relatively inexpensive from a capital investment standpoint. They also allow for the introduction of any scrubbing liquid or chemical. This feature is important because the choice of the proper chemical for odor destruction may require the use of a series of different compounds, eliminating ineffective ones until the chemical additive providing effective odor control is found.

Beside chemical oxidation, a number of odorous compounds can be altered, forming nonodorous compounds or forming other nonoffensive odorous compounds by the addition of nonoxidative chemicals. One technique often used is the addition of caustic soda or a mild acid to

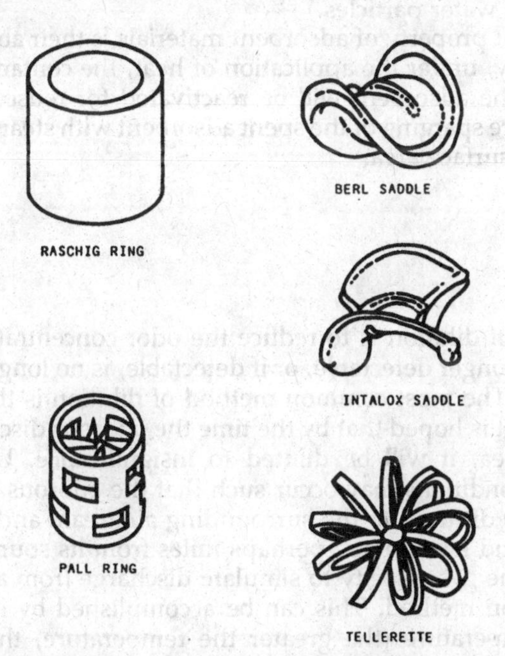

Figure 7–2 Common tower packing materials.

scrubbing water. This changes the chemical nature of the odor, not necessarily oxidizing it, resulting in an acceptable discharge. Of course, the effectiveness of this procedure is a function of the nature of the odor, its constituent analysis, and characteristics.

Adsorption

As discussed in chapter 4 "Particulate Pollutants," a number of materials have the physical property of high adsorbtive capacity, the ability to adsorb or attract gases to their surfaces. Adsorbtive media commonly used include activated carbon, silica gel, aluminum oxide, and magnesium silicate. The particles of each of these materials have extremely high area to weight ratios providing high retentive capacity.

Activated carbon is the most widely used adsorbent. One reason for its popularity is that it has a low affinity for moisture. Moisture will not compete with contaminants in searching out the carbon surface. Other adsorbents will attract moisture and will have, therefore, a short useful life in wet gas environments. These other adsorbents can lose their effectiveness by picking up the moisture present in the air as humidity over a period of time. (Note that moisture particles present in a gaseous stream will reduce the effectiveness of a carbon filter to absorb gaseous organic contaminants because of the physical blockage caused by the relatively large water particles.)

An important property of adsorbent materials is their ability to regenerate. Normally, under the application of heat, the contaminant will be released and the adsorbent will be reactivated for reuse. Reactivation may also require sparging of the spent adsorbent with steam to maximize its particulate surface area.

Dilution

The principle of dilution is to reduce the odor concentration to a level where it is no longer detectable, or if detectable, is no longer considered objectionable. The most common method of dilution is the installation of a tall stack. It is hoped that by the time the odorous discharge reaches a populous area, it will be diluted to insignificance. Unfortunately, atmospheric conditions may occur such that the odorous emissions are not adequately diffused to the surrounding airstream and the odor can return to ground level intact, perhaps miles from its source.

Increasing the gas velocity to simulate discharge from a high stack is another dilution method. This can be accomplished by increasing the gas outlet temperature (the greater the temperature, the greater the volumetric flow and the higher the resultant velocity exiting the stack, compared to a lower temperature discharge from the same stack) or by decreasing the stack diameter (decreasing the cross section of the discharge, which results in an increase in velocity). These techniques, however, are subject to the same uncertainties and limitations as are tall stacks.

Another method of odor control by dilution is the addition of clean air to the odorous gas prior to its exit from the stack. Clean air can be added by providing a flue damper immediately upstream of the induced draft fan, or, if there is no outlet fan, by use of a barometric damper at the base of the stack. These are more positive techniques of odor control than increasing the effective height of a stack, and may be used as temporary measures to reduce the immediate impact of an odorous discharge, but it is not a reliable means of odor control over the long term.

Masking

Attempts are often made to counteract, or neutralize, an odorous discharge by the addition of odorous nonreactive vapors. Ideally, the odor intensity may be decreased to zero, a condition known as cancellation. This can rarely occur. At best, the resultant of two odors will be a more pleasant odor. In the worst case, the combination of two odors can be a less desirable odor.

Firms specializing in this form of odor control employ experts who are capable of choosing an appropriate masking agent based on the quality of the odor in question. This work approaches the status of an "art" since it involves more than simply practical applications of science or technology.

Effective Control

If a process cannot be altered to effectively eliminate the generation of objectionable odors, other techniques must be considered. The most effective of these methods of odor control are direct-flame and catalytic incineration, but these are also the more expensive methods. Dilution is the least effective of these techniques; however, they are usually the least costly. Masking is often applied as a relatively low-cost technique and it can be more effective than dilution. Adsorption has less application than other systems. Absorption is an effective means of odor control for many applications, although it is not as effective as incineration. Both sorbtive techniques are significantly lower in cost than incineration.

References and Bibliography

1. Zwaardemaker, H. 1877. Die physiologie der geruchs, Leipzig. W. Engelmann.
2. Horstman, S. Identification of community odor problems by use of an observer corps. *Journal of the Air Pollution Control Association* (November 15, 1965).
3. McCord, C., and W. Witheridge. 1949. *Odors, physiology and control.* New York: McGraw Hill.
4. Cross, F. 1973. *Air pollution odor control primer.* TECHNOMIC.
5. Dravnieks, A. Odor perception and odorous air pollution. *TAPPI* (May 1972) 55 5:737–742.
6. Dravnieks, A. Measurement of sensory properties of odors. *ASTM Standardization News* (March 1983).
7. Standard test methods for measurement of odor in atmospheres. *ANSI/ASTM D1391*, Phildelphia, PA, 1978.
8. Sawyer, C., and P. Kahn. Temperature requirements for odor destruction in sludge incineration. *WPCF Journal* (December 1960): 1274–1278.
9.. USEPA. 1980. *Regulatory options for the control of odors*, EPA 450/5-80-003. Washington, DC: Government Printing Office, February.

8

Incineration

Incineration is the destruction of waste materials by the controlled application of heat. There are a wide variety of incinerators that have been developed for the burning of the varied types and quantities of wastes found in today's society. Specialty incinerators have been developed for radioactive wastes, off-the-shelf units for hospital or apartment house waste, incineration equipment for burning ordinary household refuse, industrial incinerators for the destruction of liquid waste, and so on. Some of these incinerators have heat recovery equipment that generates steam, hot water, or hot air, and some do nothing more than burn their designated waste streams. There are incinerators that are versatile in their application, able to burn anything from waste gas to garbage, tars to waterborne wastes, all within one unit. This chapter describes some of the various incinerators available for waste destruction.

Single Chamber Incinerators

A typical single-chamber incinerator is shown in Figure 8–1. It is used for the destruction of solid waste. The charge is placed on the grate and is fired. In its simplest form, it is fired manually, with a match, and the flue gas generated flows up the stack and into the atmosphere. Air is naturally induced to flow within the chamber through the under- and overfire air ports as shown. The air entering below the grate, underfire air, provides a source of oxygen to burn the waste. The overfire air supply promotes the burning of the flue gases, which will be rich in unburned carbon and hydrocarbon gases and particles. This is a simple batch-charged unit. A load is charged, is fired, and then is allowed to burn out, perhaps overnight. The resultant ash falls to the bottom of the chamber and is removed by hand at the conclusion of the burn.

Variations of this incinerator have automatic charging, and may be provided with a burner that can fire auxiliary fuel to burn the waste, if the waste material does not have sufficient heat value to support its own combustion. Some types of single-chamber incinerators will have afterburners placed within the stack, immediately upstream of the burning chamber. Use of the afterburner will help destroy unburned combustibles within the off-gas, producing a significantly cleaner exhaust to the atmosphere.

The flue-fed incinerator, shown in Figure 8–2, is another type of single-chamber incinerator. It has had widespread use for incinerating domestic

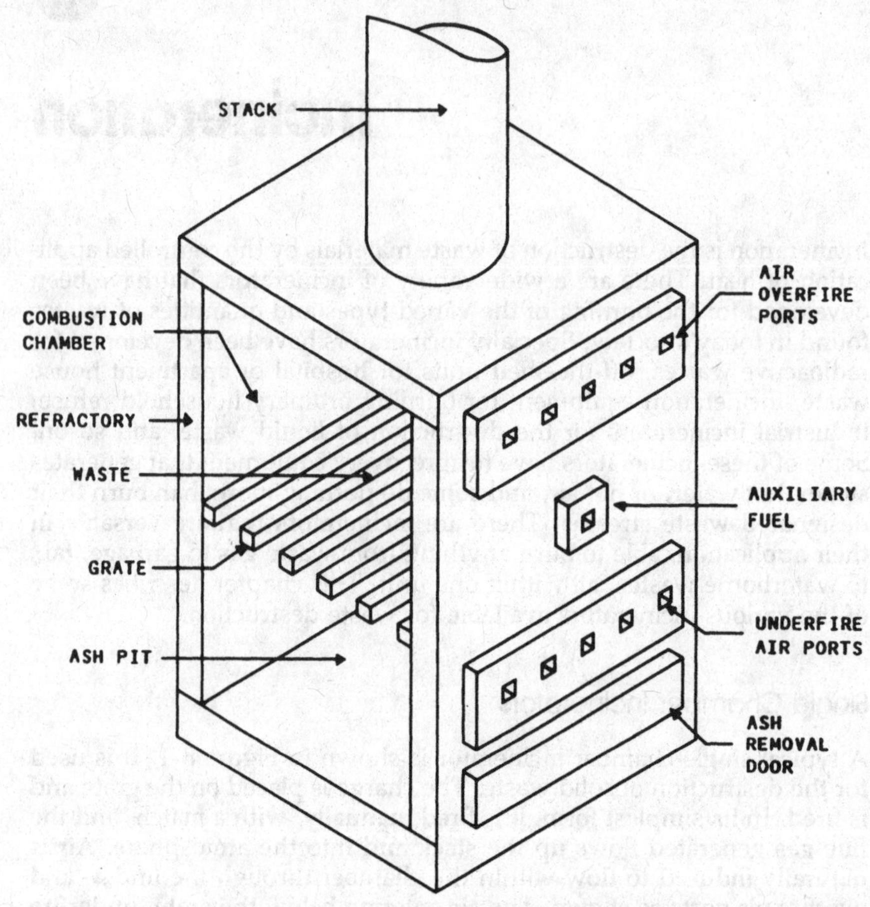

Figure 8–1 Single-chamber incinerator.

solid waste in apartment buildings. This illustration shows an incinerator with two burners, one for igniting the waste and another, acting as an afterburner, to cleanse the flue gas prior to its discharge. The draft control damper is used to control the flow of air into the burning chamber. Opening this damper will allow a greater airflow into the system. With too much air, the emissions will increase with excessive carryover of solid material from the grate into the gas stream. Also, the excessive air will tend to cool the burning chamber, preventing effective burnout of the waste and the waste's offgas. On the other hand, insufficient airflow will prevent effective combustion of the waste. Instead of burning, the waste material will smolder, with incomplete combustion taking place. The products of incomplete combustion are extremely dirty gases and an unburned "ash" which is subject to further burning.

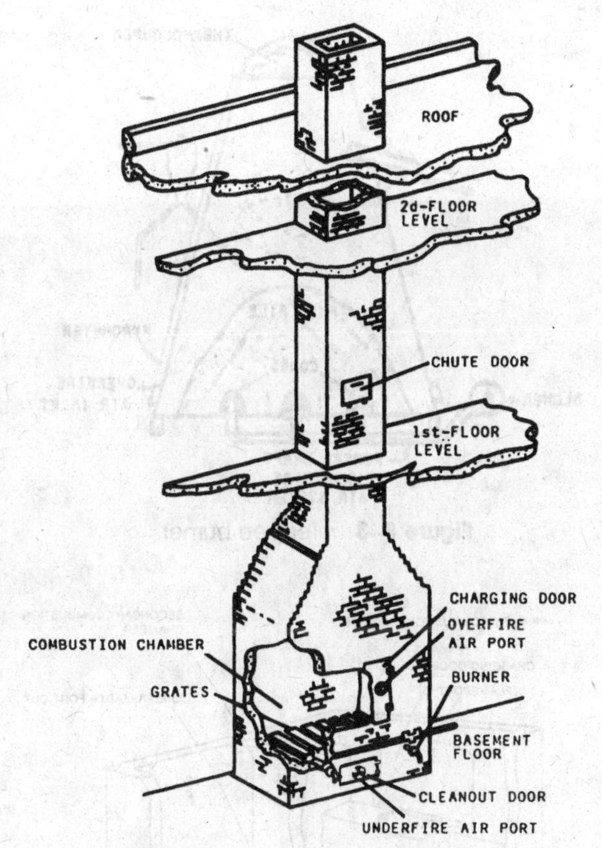

Figure 8–2 Flue-fed incinerator.

Teepee Burners

These incinerators, also known as wigwam and conical incinerators, had been extremely popular for the destruction of refuse, wood, and agricultural waste until the advent of strict air pollution control regulations. They are conical steel shells as shown in Figure 8–3. Waste is usually dropped from an entrance near the top of the cone to a series of cones elevated from the "teepee" floor. Flue gas discharges directly from the teepee or conical top without provision of a stack or other exiting gas flow control equipment.

Multiple-chamber Incinerators

In an attempt to provide complete burnout of combustion products and decrease the airborne particulate loading in the exiting flue gas, multiple chamber incinerators have been developed. A first, or primary chamber

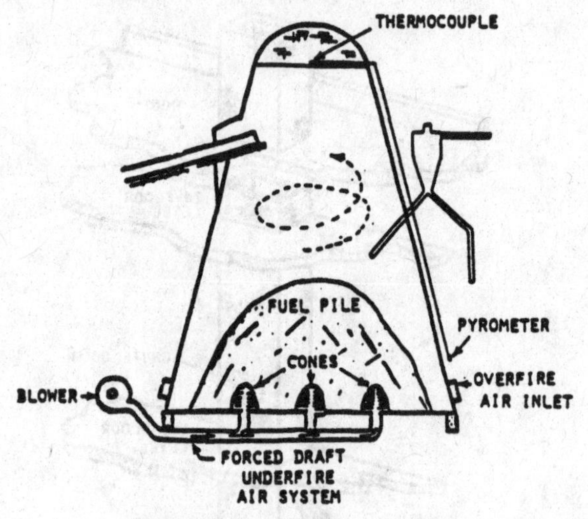

Figure 8–3 Teepee burner.

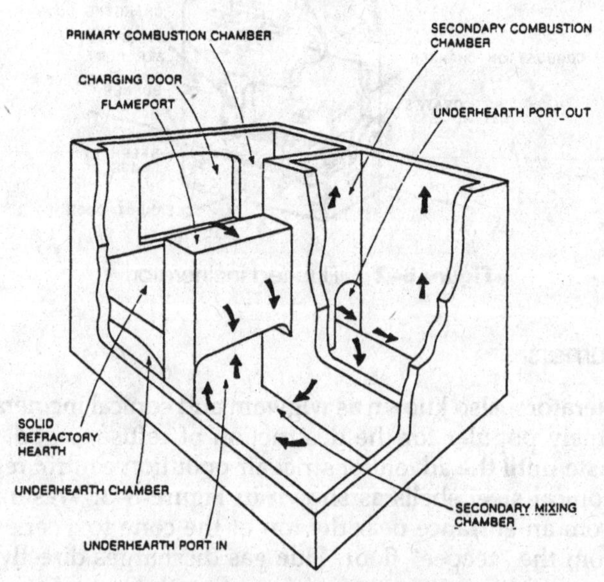

Figure 8–4 Retort incinerator.

is used for combustion of solid waste. The secondary chamber provides the residence time, and supplementary fuel, for combustion of the unburned gaseous products and airborne combustible solids (soot) discharged from the primary chamber.

There are two basic types of multiple chamber incinerators, the retort and the in-line systems.

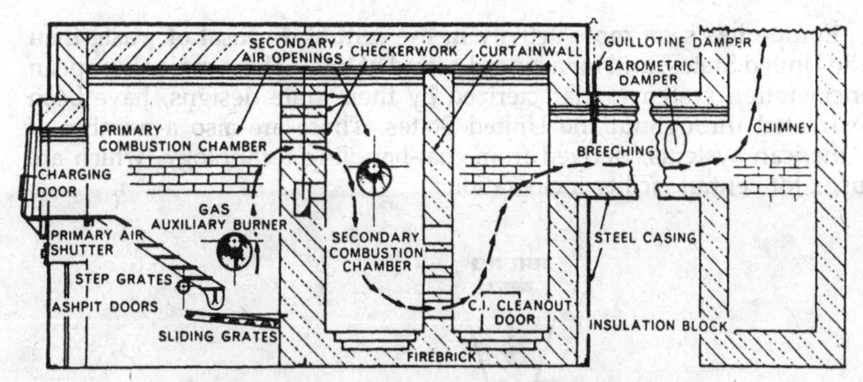

Figure 8–5 Gas-fired in-line incinerator.

- Retort Incinerator: This unit is a compact cube-type incinerator with multiple internal baffles. The baffles are positioned to guide the combustion gases through 90° turns in both lateral (horizontal) and vertical directions. At each turn ash (soot) drops out of the flue gas flow. The primary chamber has elevated grates for burning of the waste and an ash pit for collection of ash residual. A cutaway view of a typical retort incinerator is shown in Figure 8–4.
- In-Line Incinerator: This is a larger unit than the retort incinerator. Flow of combustion gases is straight through the incinerator, axially, with abrupt changes in direction as shown in Figure 8–5. Waste is charged on the grate, which can be either stationary or moving. As with the retort type, changes in the flow path and flow restrictions in an in-line incinerator provide settling out of larger airborne particles and increase turbulence for more efficient burning. Supplemental fuel burners in the primary chamber ignite the fuel whereas secondary chamber supplementary fuel burners provide heat to maintain the temperature required for complete combustion of the unburned components of the exhaust gas.

Central Disposal Incineration Systems

The term central waste disposal refers to the process of routing waste from many sources to a centrally located single facility for disposal, usually by incineration. Municipalities utilize central disposal for household waste and many industries will send their waste products to a central facility for incineration, particularly when an industry has many separate plants in relatively close proximity to each other. The economics of scale can often justify provision of heat recovery equipment, usually in the form of hot water or steam generation, from central disposal systems, whereas it may not be economically justifiable from smaller facilities.

Europe has seen more activity in the central disposal of waste than the United States, and for a longer period of time. Three major European incineration systems, characterized by their grate designs, have been marketed throughout the United States. There are also a number of American systems, derived from coal-handling technology, which are used for central facility incineration.

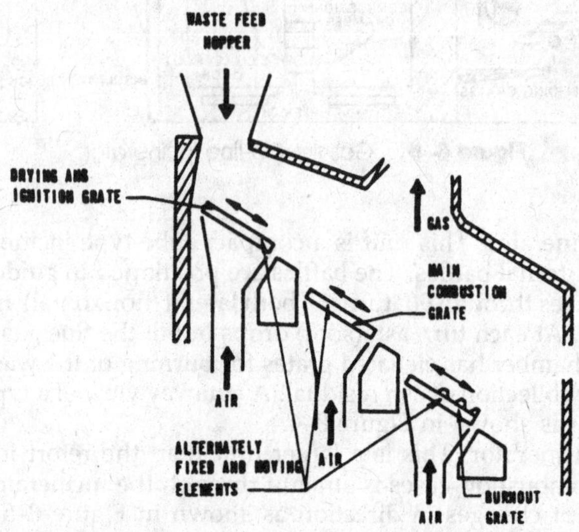

Figure 8–6 Reciprocating grate system. *Source: Wheelabrator-Frye, Inc., [Von Roll System])*

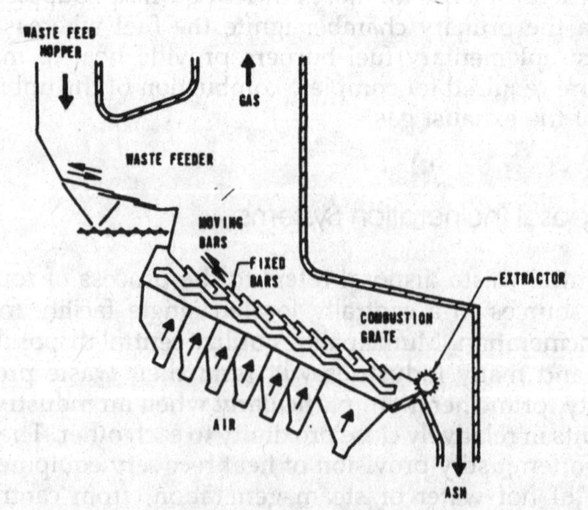

Figure 8–7 Reverse reciprocating grate system. (*Source: Ogden-Martin, Los Angeles, CA*)

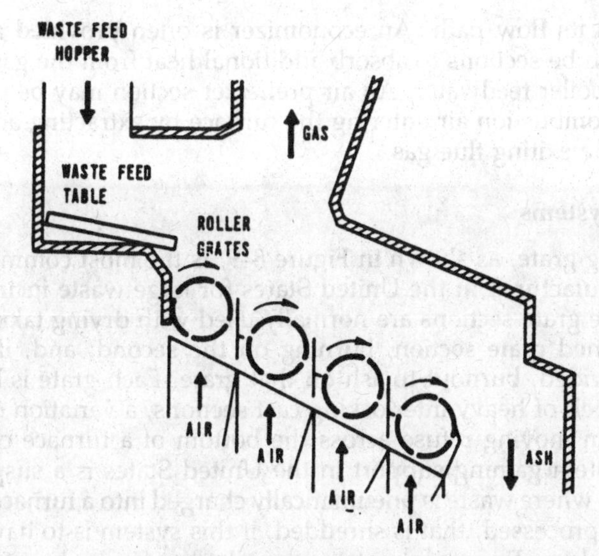

Figure 8–8 Drum grate system. (*Source:* Brown and Ferris Industries [VKW System])

European Systems

The three predominant incineration systems developed by European manufacturers are the reciprocating grate, roller (drum) grate, and reverse reciprocating grate systems. The furnace shown in Figure 8–6 has a reciprocating grate system. Refuse dries out on the first grate section, burns on the second series of grates, and burns out to ash on the third grate. Adjacent grate elements are fixed and moving. The action of the moving grates drives the refuse or ash forward to a stationary section. Material moves from the fixed grate to the adjacent moving grate by the action of new material entering the fixed grate zone. This successive movement results in a semicontinuous forward motion of material toward the end or bottom of the furnace while providing turbulence of the waste to encourage efficient burning.

A system of reverse reciprocating grates is shown in Figure 8–7. As the grates move forward, and then reverse, the waste is subject to continual agitation and forward motion, down toward an extractor section that aids in the discharge of ash from the furnace chamber.

Large hollow roller or drum grates are used in the system shown in Figure 8–8. The drums rotate slowly, gently agitating the waste and moving it along to subsequent drums. As in the other grate systems, air is provided beneath the grates for grate cooling and also as a supply of combustion air (underfire air) for the waste. The furnace walls are normally built of finned water tubes welded together, absorbing some of the furnace heat. The flue gases will exit the furnace chamber at approximately 1800°F and will pass through a section of vertical boiler tubes across its flow path. An economizer is often provided after the main boiler tube sections to absorb additional heat from the gas stream to heat the boiler feedwater. An air preheater section may be provided to preheat combustion air entering the furnace by extracting additional heat from the exiting flue gas.

American Systems

The traveling grate, as shown in Figure 8–9, is the most common grate system manufactured in the United States for large waste incinerators. Two or three grate sections are normally used with drying taking place on the inclined grate section, burning on the second, and, if a third grate is provided, burnout to ash on that grate. Each grate is basically an endless belt of heavy interlocking cast sections, a variation of a conveyor system moving refuse across the bottom of a furnace chamber. Another system gaining support in the United States is a suspension-fired system where waste is pneumatically charged into a furnace. Waste must be preprocessed, that is shredded, if this system is to have application. The above European systems are all mass fired where no waste processing is necessary. The refuse hoppers, grates, and ash disposal systems are all designed to handle bulk, heterogenous waste as received. Blowing refuse into a furnace so that most of the burning occurs within

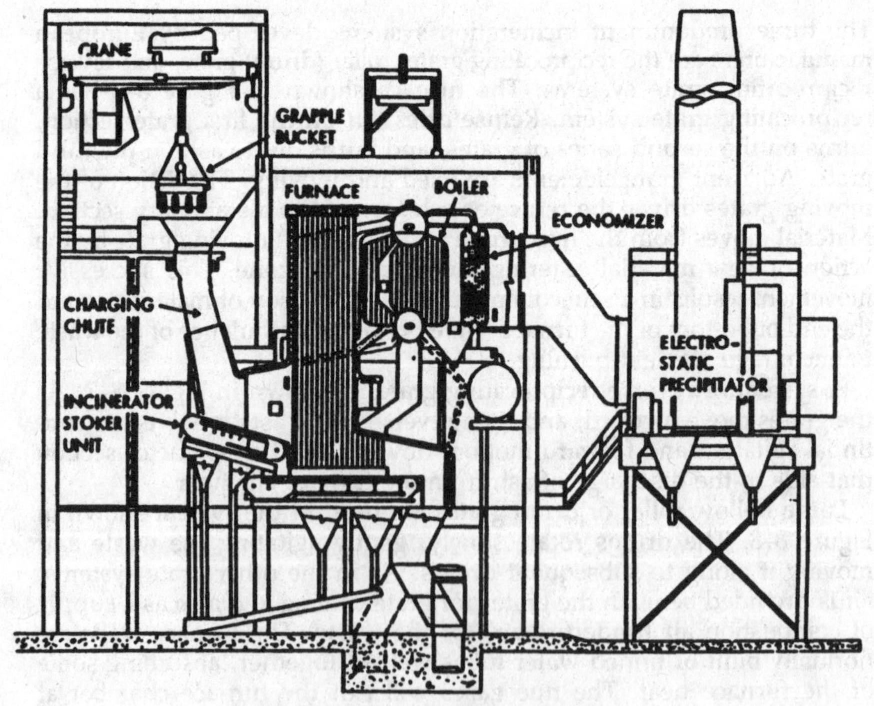

Figure 8–9 Traveling grate system.

the furnace volume, and not on a grate, requires that the waste be shredded to relatively small mean size. Suspension fired systems normally have a single or duplex traveling grate system to catch those waste materials that have not been burned in suspension and to remove the ash generated from within the furnace.

The American as well as the European systems are normally built with water-wall construction and other boiler tube sections to maximize the generation of steam from the incinerator flue gas flow.

Pyrolysis

There are a number of systems which have been developed using pyrolysis for waste destruction, a process related to incineration. Incineration requires that sufficient air (oxygen) be provided to adequately combust the waste to its basic combustion products, primarily carbon

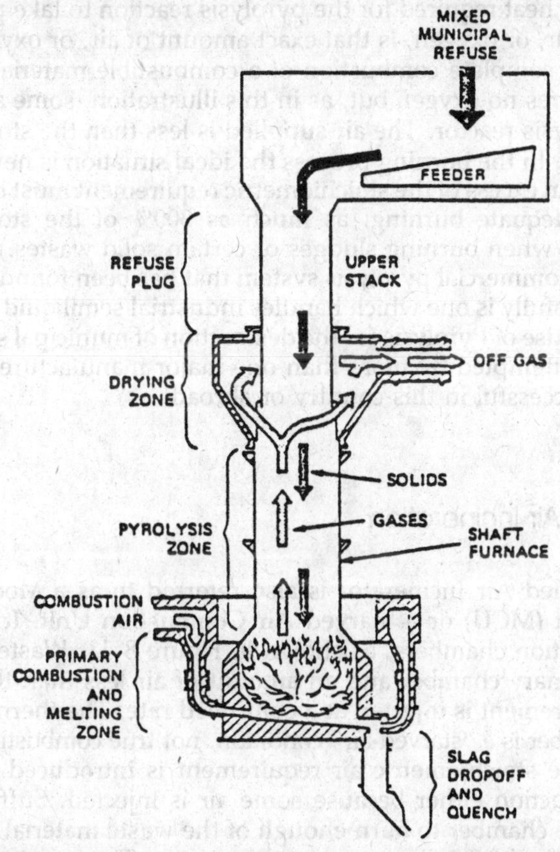

Figure 8–10 Pyrolysis System. (*Source: Andco-Torrax*)

dioxide and water vapor. Pyrolysis is the destructive distillation of a solid material in the presence of heat and in the absence of air, or oxygen. It is an endothermic reaction; that is, heat must be provided for the reaction to occur, whereas incineration will in itself produce heat (exothermic).

A typical pyrolytic reaction is as follows, utilizing the chemical formula of cellulose, the main constituent of paper:

$$C_6H_{10}O_5 \rightarrow CH_4 + 2\ CO + 3\ H_2O + 3\ C$$

A gas is produced containing methane (CH_4) carbon monoxide, (CO) and moisture. The carbon monoxide and methane are combustible, providing the off-gas with a positive heating value. The carbon residual (3C) is contained within the furnace solid residual, which also contains metals, metallic oxides, oxides, and other minerals.

Three major systems utilizing a variation of the pyrolysis process have been marketed in the United States. A typical system is shown in Figure 8–10. Air is injected into the furnace to provide just enough burning to generate the heat required for the pyrolysis reaction to take place. (Stoichiometric air, or oxygen, is that exact amount of air, or oxygen, that is required for complete combustion of a combustible material. Pyrolysis ideally requires no oxygen but, as in this illustration, some air is added to the pyrolysis reactor. The air supplied is less than the stoichiometric requirement. In the burning process the ideal situation is never reached and air well in excess of the stoichiometric requirement must be provided to insure adequate burning, as much as 300% of the stoichiometric requirement when burning sludges or certain solid wastes.)

The only commercial pyrolysis system that has been found to be operating successfully is one which handles industrial semiliquid and sludge wastes. The use of Pyrolysis for the destruction of municipal solid wastes have been attempted by more than one major manufacturer, but have not been successful in this country or abroad.

Controlled Air Incineration

The Controlled Air Incinerator is also referred to as a Modular Combustion Unit (MCU) or a Starved Air Combustion Unit. It consists of two combustion chambers, as shown in Figure 8–11. Waste is charged into the primary chamber and an amount of air less than the stoichiometric requirement is injected at a controlled rate. The thermal reaction in this chamber is a "starved-air" condition, not true combustion, because less than the stoichiometric air requirement is introduced. It is not a pyrolysis reaction either because some air is injected. Sufficient air is added to the chamber to burn enough of the waste material to generate the heat required for maintenance of the process. The temperature within

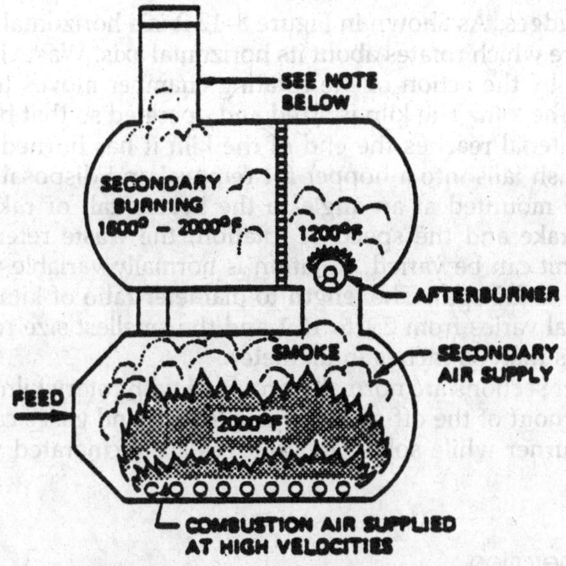

Figure 8–11 Controlled air incinerator.

the primary chamber is usually maintained in the range of 1400°F to 1600°F.

Air is injected into the secondary chamber for combustion of the off-gas, which contains combustible organic material released from the waste within the primary chamber. Temperatures in the secondary combustion chamber can be as high as 2200°F. With paper-type wastes the secondary combustion chamber temperature will normally be held in the range of 1600°F to 1800°F. With these high temperatures and with the large volume within the combustion chamber, providing retention time for the hot gases, the exiting flow is relatively clean.

The small airflow injected into the primary chamber, where the waste is charged, results in less turbulence and therefore less carryover of particulate matter into the gas stream than can be expected from incineration processes operating with air in excess of stoichiometric requirements. Another feature of the starved air incinerator is the simplicity of its control. Air is introduced in response to the temperature in each chamber and this control is usually automated. The only other means of process control normally employed is regulation of the quantity of waste fed to the unit.

Rotary Kiln

This incinerator is the most flexible and versatile of the incinerators in common use today. It is widely used for the destruction of industrial wastes and has had application in the incineration of municipal solid waste and sludges. As shown in Figure 8–12 it is a horizontal refractory lined structure which rotates about its horizontal axis. Waste is fed from one end and by the action of the rotating chamber moves toward the other end of the kiln. The kiln is sized and operated so that by the time the waste material reaches the end of the kiln it has burned out to an ash and the ash falls into a hopper for removal and disposal.

The kiln is mounted at an angle to the horizontal, or rake, and by varying the rake and the speed of rotation, the waste retention time within the unit can be varied. Rotation is normally variable within the range of 0.25 to 2.5 rpm. The length to diameter ratio of kilns used for waste disposal varies from 2:1 to 10:1 and the smallest size rotary kilns rarely are less than 36 inches in diameter.

Afterburner sections are normally provided with rotary kilns to insure complete burnout of the off-gases. Waste liquids and gases can be fired in the afterburner while solids and sludges are incinerated within the kiln itself.

Sludge Incineration

Sludges are generally defined as small solid particulate matter suspended in water or in another liquid. The total solids content is at least 15% and can be as high as 60%. A number of incinerator systems have been developed specifically for the destruction of sludge material.

Multiple Hearth Incinerator

This is a vertical cylindrical structure, refractory lined, with a series of refractory hearths positioned one beneath the other, as shown in Figure 8–13. The center shaft, hollow to allow the passage of cooling air through

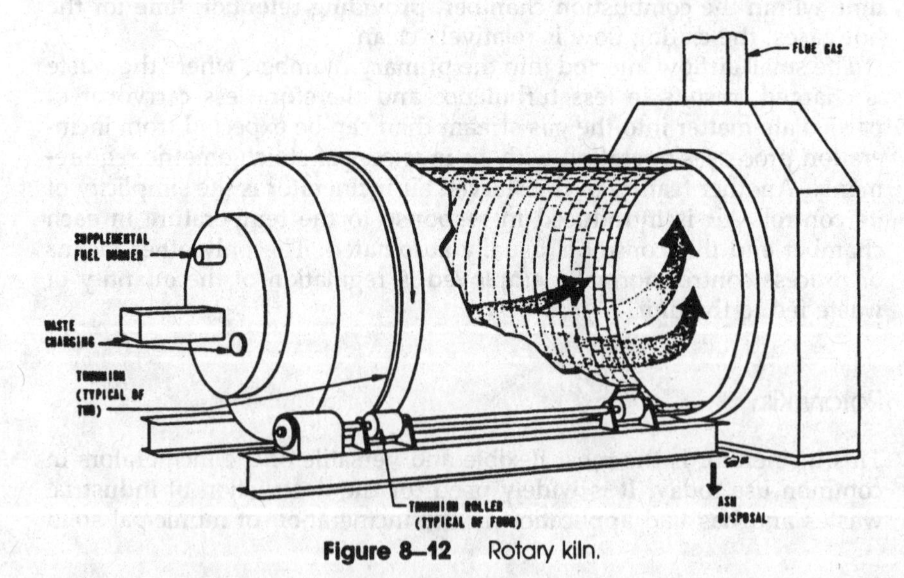

Figure 8–12 Rotary kiln.

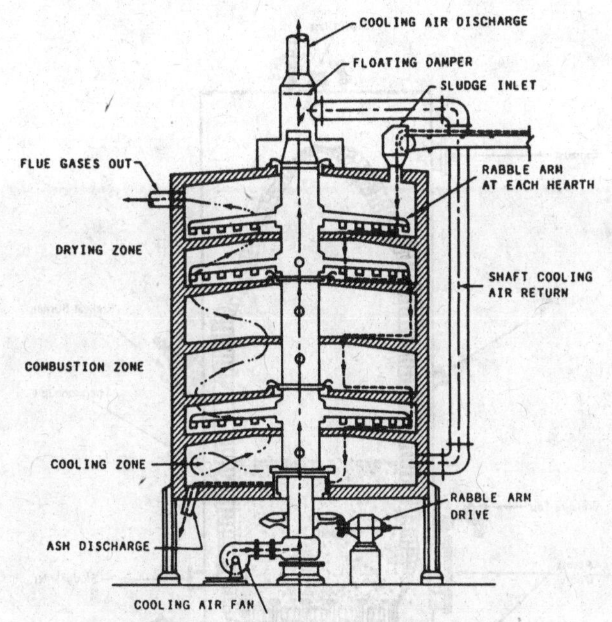

Figure 8–13 Multiple-hearth incinerator.

it, rotates within the incinerator at approximately one revolution per minute carrying the rabble arms with it, sweeping sludge across the hearths. Sludge is rabbled from the edge of one hearth to its center, and drops to the next hearth, where it is rabbled to the outside of that hearth, and so on. Alternate hearths have large center openings, or outside drop holes, through which the sludge falls.

Sludge is dried on the top hearths and starts to burn toward the center of the furnace. It burns out to ash at the bottom of the furnace where the ash is discharged.

These furnaces range from 6 to 26 feet in diameter and can normally have from 5 to 12 hearths. Gas outlet temperatures, which range from 800°F to 1200°F when burning organic sludges, is approximately 600°F lower in temperature than the highest temperature within the furnace, 1400°F to 1800°F within the combustion zone.

Fluid Bed Incinerator

As shown in Figure 8–14, the fluid bed incinerator is a steel refractory lined chamber with a bed supporting plate separating the windbox from the reactor, and supporting a bed of sand. Air is heated and injected into the sand bed. At a temperature of 600°F the air will generate sufficient turbulence within the sand bed to produce fluidization; that is, the sand will appear to act as a fluid. Sludge is injected within the reactor and as soon as it contacts the turbulent sand/air bed, water flashes from the sludge particle and oxidation begins.

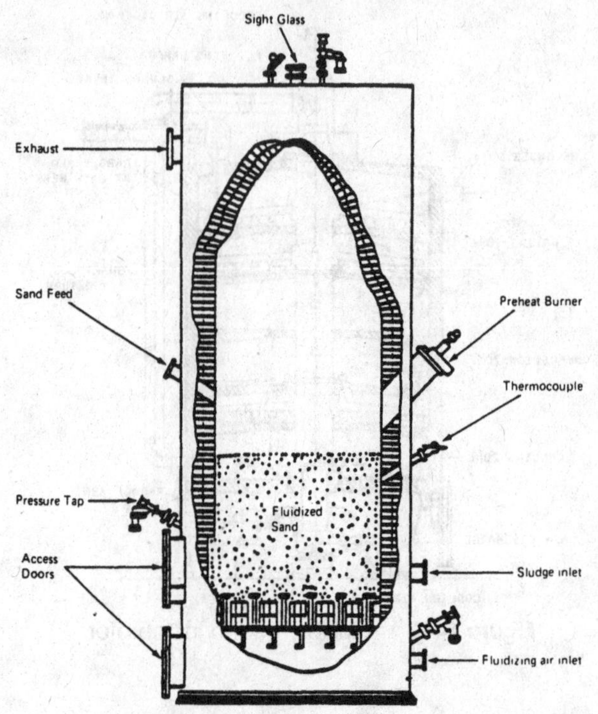

Figure 8–14 Fluid bed incinerator.

Ash residual from sludge combustion is airborne and exits from the reactor airborne in the flue gas it is removed from the gas stream by high-energy scrubbing equipment downstream of the incinerator exit. The reactor exit temperature is normally in the range of 1400°F to 1600°F. Fluid bed systems typically utilize an air preheater (or recuperator) at the reactor exit, heating the fluidizing air to 1000°F while reducing the exiting flue gas temperature by 500°F to 700°F.

Liquid Waste Incineration

In general, liquids can be pumped if their viscocity is less than 10,000 Saybolt Seconds, Universal (SSU). They can be fired and burned in suspension with viscosities up to 750 SSU. Specialty burning equipment is available for suspension firing liquids with viscosities up to 5,000 SSU. Liquid incinerators are those furnaces which have been designed specifically for the burning of waste liquids in suspension.

Liquid incinerators are either horizontal or vertical cylindrical chambers, normally lined with refractory material, with nozzles firing axially or tangentially into the furnace. Air or steam is usually injected into the waste nozzle helping to atomize the waste stream to increase burning

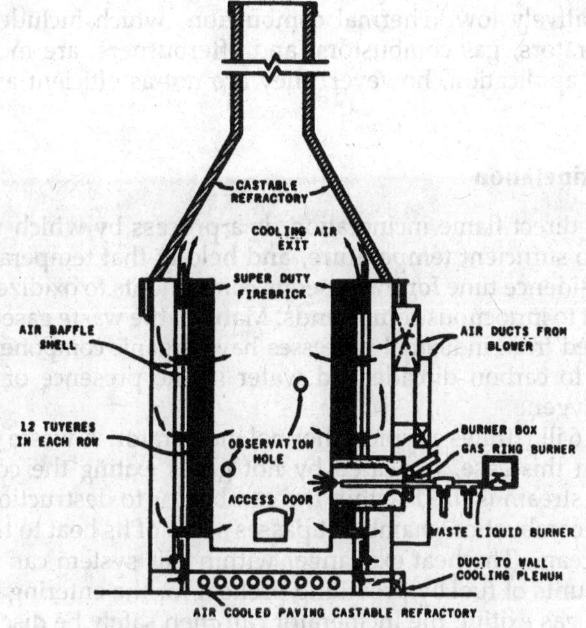

Figure 8–15 Liquid waste incinerator.

efficiency. Secondary air is provided downstream of the waste nozzle(s) in a manner that increases turbulence within the furnace which in turn promotes the burning process.

A typical radial fired vertical liquid waste incinerator is shown in Figure 8–15. Waste is injected through a burner box into the interior of the furnace. Air is introduced through the chamber wall via small openings, tuyeres, which are placed in a manner that imparts a swirling motion to the airflow, creating turbulence within the furnace. Additional air passes through the floor of the furnace and the furnace wall, acting as a coolant, before it is introduced into the burning zone higher in the furnace.

In the particular furnace illustrated, gas is the prime fuel, injected through a ring burner to heat the furnace from a cold condition and to maintain heat within the furnace when insufficient heat is generated by burning of the liquid waste.

Gaseous Waste Incineration

There are basically two means of gaseous waste destruction in common usage, thermal and catalytic incineration. Catalytic incinerators are limited in their application. They cannot handle all gases nor gasborne components and the proportion of combustible gas within the stream must be relatively low. Thermal combustion, which includes flaring, fume incinerators, gas combustors, and afterburners, are more nearly universal in application; however, they are not as efficient as catalytic incinerators.

Thermal Incineration

Thermal, or direct flame incineration, is a process by which waste gas is brought to sufficient temperature, and held at that temperature for a sufficient residence time for the gaseous components to oxidize or otherwise convert to innocuous compounds. Many of the waste gaseous materials generated from industrial processes have organic components which will reduce to carbon dioxide and water in the presence of heat and sufficient oxygen.

Figure 8–16 illustrates a typical thermal incinerator. A waste gas, process fumes in this case, is heated by hot gases exiting the combustor. This heated stream is fired within the combustor to destruction, and as it leaves the combustion chamber it passes some of its heat to the incoming fume stream. The heat exchanger within this system can save substantial amounts of fuel by providing preheat for the entering waste gas stream. The gas exiting the incinerator can then safely be discharged to the atmosphere.

Organic gases normally require a temperature in the order of 1400°F and residence time of approximately 0.5 seconds, in addition to proper

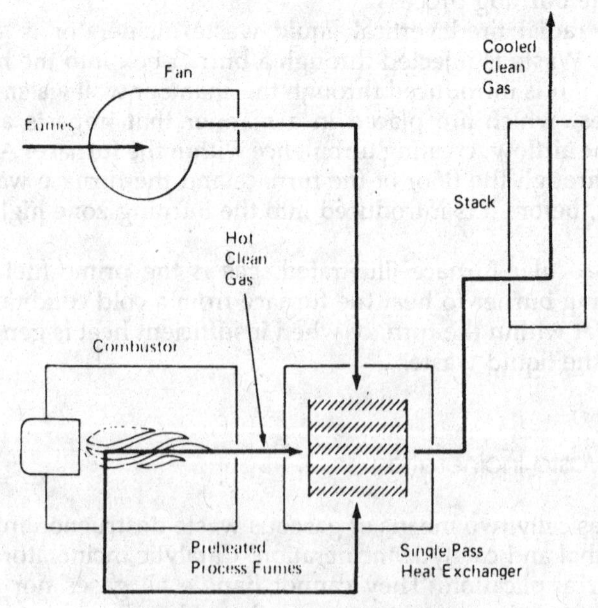

Figure 8–16 Fume incinerator.

turbulence, or mixing with air, for complete destruction of the contaminant. The heat exchanger can extract from 200°F to as much as 1000°F from the exhaust stream.

Catalytic Incineration

This is a method of destruction of organic components utilizing a catalyst material to allow oxidation to occur at lower temperatures than direct combustion requires. For instance, benzene requires 1076°F for burning to occur in air; however, using a catalyst the temperature required for the reaction is 575°F. This temperature difference represents a savings of 67% of the energy requirement for combustion of low concentrations of benzene when using catalytic incineration. Similarly, the utilization of catalytic incineration can save 86% of the energy cost when burning ethanol, 60% when burning carbon monoxide, and 70% when burning toluene, all in small quantities. (When high concentrations of these materials are present in a gas stream they will burn without the addition of supplemental fuel and, therefore, catalytic incineration would not be applicable.)

Catalytic incinerators will often utilize heat exchangers to provide even greater savings in fuel consumption, as shown in Figure 8–17. The incoming gas stream is heated before entering the furnace chamber by the hot gas exiting the chamber.

Catalyst materials normally used are the noble metals, platinum, palladium, or rhodium. Other materials which act as catalytic agents are

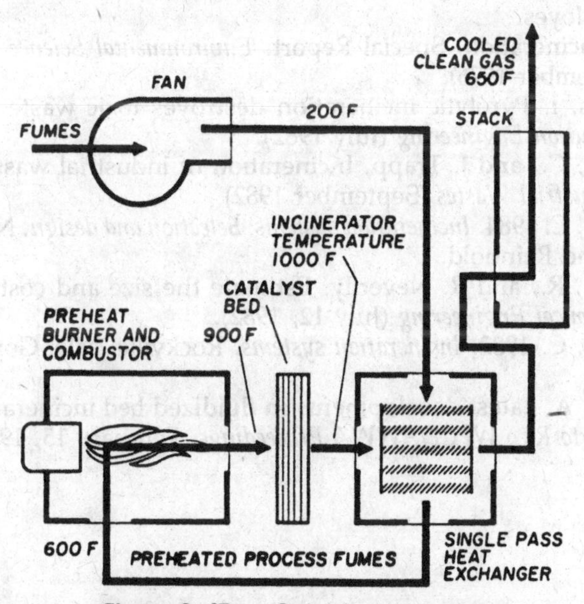

Figure 8–17. Catalytic incinerator.

copper chromite and the oxides of copper, chromium, manganese, nickel, and cobalt. Stainless steel mesh or wire, or porcelain structures, normally supported on alumina struts, are used as the matrix for deposition of the catalyst material.

References and Bibliography

1. USEPA. 1974. *St. Louis: Union electric refuse firing demonstration air pollution test report. EPA 650/2-74-073*. Washington, DC: Government Printing Office, August.
2. Fluidized bed combustion. *Compressed Air* (September 1981).
3. Berkau, E. Current research on the treatment and destruction of industrial wastes. *EPA Hazardous Waste Conference Proceedings*, Ft. Mitchell, KY, March 1982.
4. Frankel, I., and N. Sanders. Profile of the hazardous waste incinerator manufacturing industry. *EPA Hazardous Waste Conference Proceedings*, Ft. Mitchell, KY, March 1982.
5. Chapman, G., and R. Johnson. Shipboard incineration of hazardous chemical waste. *EPA Hazardous Waste Conference Proceedings*, Ft. Mitchell, KY, March 1982.
6. Sittig, M. 1979. *Incineration of Industrial Hazardous Wastes and Sludges*. Park Ridge, NJ.: Noyes.
7. Dillion, A. 1981. *Hazardous Wastes Incineration Engineering*. Park Ridge, NJ: Noyes.
8. Brunner, C. 1980. *Design of Sewage Sludge Incineration Systems*. Park Ridge, NJ: Noyes.
9. Sludge incineration. Special Report. *Environmental Science and Technology* (November 1976).
10. Williams, I. Pyrolytic incineration destroyes toxic waste, recovers energy. *Pollution Engineering* (July 1982).
11. Brunner, C., and J. Trapp. Incineration of industrial waste in Cincinnatti. *Industrial Wastes* (September 1982).
12. Brunner, C. 1984. *Incineration systems: Selection and design*. New York: Van Nostrand Reinhold.
13. Vatavuk, R., and R. Neverily. Estimate the size and cost of incinerators. *Chemical Engineering* (July 12, 1982).
14. Brunner, C. 1982. *Incineration systems*. Rockville, MD: Government Institutes.
15. Baturay, A. Latest developments in fluidized bed incineration technology, *Puerto Rico WPCF/AWWA Proceedings*, February 15, 1984.

9

Incinerator Calculations

Incinerator calculations cannot be exact because of the nature of the materials normally incinerated. Almost by definition, waste is nonuniform in quality and will vary in composition from lot to lot. The measurement of waste characteristics is in itself a difficult, if not impossible task. How can a reliable determination of heating value be made for municipal solid waste for instance, when one sample may be over 90% cardboard with no metals while another sample contains plastic, glass, and metal containers? The heterogeneous nature of most waste streams lead to inaccuracies and uncertainties in too stringent an analysis.

Past experience is the best guide in incinerator design and in the determination of incinerator characteristics. In this chapter empirical data will be discussed as well as a method for calculating incinerator discharges based on waste composition.

Paper Waste

The majority of municipal solid waste is paper, cardboard, and related items. The prime constituent of these wastes is cellulose, $C_6H_{10}O_5$, which accounts for over 70% of the structure of wood. By examining the combustion properties of cellulose, therefore, burning characteristics of paper and similar waste materials can be evaluated.

Combustion of Cellulose

Cellulose will burn as follows:

$$\begin{array}{lllll} C_6H_{10}O_5 & + \ 6\ O_2 & \rightarrow 6\ CO_2 & + \ 5\ H_2O & \\ 162.16 & 192.00 & \rightarrow 264.06 & 90.10 & \\ 1.00 & 1.18 & 1.63 & 0.55 & \end{array} \qquad (9\text{-}1)$$

The first row beneath the equilibrium equation is the atomic weight of the compound above multiplied by the number of molecules, or moles, present. The individual atomic weights are carbon, 12.01, hydrogen,

Table 9–1 Air Composition, Fraction

Component	By Weight	By Volume
Oxygen in Air	0.2315	0.21
Nitrogen in Air	0.7685	0.79
Air to Oxygen	4.3197	4.7619
Air to Nitrogen	1.3012	1.2658
Oxygen to Nitrogen	0.3012	0.2658
Nitrogen to Oxygen	3.3197	3.7619
Molecular Weight	28.914	–

1.01, oxygen, 16.00, and nitrogen, 14.01. The second row is the atomic weight divided by the atomic weight of cellulose. These figures are the oxygen required and combustion products produced per pound of cellulose burned:

- 1.18 pound oxygen required per pound of cellulose
- 1.63 pound carbon dioxide produced per pound of cellulose
- 0.55 pound water vapor produced per pound of cellulose

Normally, air supplies the oxygen required for combustion. Table 9–1 lists the weight and volume fractions of oxygen and nitrogen in air, its two major components. The oxygen required for combustion, 1.18 lb, will carry with it 3.3197 lb of nitrogen for each lb of oxygen from the air. Therefore, for each pound of cellulose burned:

$$1.18 \cdot 3.3197 = 3.92 \text{ lb of nitrogen present.}$$

Nitrogen will not react with cellulose or other elements present to any significant degree when considering the mass balance. It will pass through the reaction without change, as nitrogen, exiting in the exhaust. Including nitrogen in the input and output flows, on a weight basis per pound of cellulose:

	In	Out
Cellulose	1.00	none
Oxygen	1.18	none
Nitrogen	3.92	3.92
Carbon Dioxide	none	1.63
Water Vapor	none	0.55
Total	6.10	6.10

The air required for combustion is 1.18 + 3.92 = 5.10 pounds per pound of cellulose: the dry gas produced is 3.92 + 1.63 = 5.55 pounds per pound of cellulose and the water vapor produced is 0.55 pound per pound of cellulose. The equilibrium equation for the combustion of cellulose, including nitrogen, is as follows:

$$\begin{array}{cccccc} C_6H_{10}O_5 + & 6\ O_2 + & 22.56\ N_2 \rightarrow & 6\ CO_2 + & 22.56\ N_2 + & 5\ H_2O \\ 1.00 & 1.18 & 3.92 & 1.63 & 3.92 & 0.55 \end{array} \quad (9\text{-}2)$$

This equation represents stoichiometric combustion which is that idealized reaction where only sufficient oxygen (or air) is provided for complete combustion, no more and no less.

Excess Air Combustion

As an ideal reaction, stoichiometric combustion is not achieveable with real equipment. Combustion is a function of waste size, air mixing, temperature, residence time, furnace geometry, and many other factors. To insure that complete combustion occurs, air in excess of the stoichiometric requirement must be provided. The total air supplied to a reaction is normally referred to in terms of the stoichiometric air requirement. For instance, when 50% excess air is called for, this requirement is 50% of the stoichiometric air supply plus the basic stoichiometric demand. In total, 150% of the stoichiometric demand is required.

Burning cellulose with 50% excess air (150% total air) the oxygen requirement is 50% in excess of the stoichiometric requirement; that is, 6 moles stoichiometric + 0.5 · 6 moles = 9 moles total air. The equilibrium equation is as follows:

$$\begin{array}{cccccccc} C_6H_{10}O_5 & 9\ O_2 & + 33.86\ N_2 & 6\ CO_2 & + 33.86\ N_2 & + 5\ H_2O & + 3O_2 \\ 162.16 & 288.00 & 948.76 \rightarrow & 264.06 & 948.76 & 90.10 & 96.00 \\ 1.00 & 1.78 & 5.85 & 1.63 & 5.85 & 0.56 & 0.59 \end{array} \quad (9\text{-}3)$$

Note that the excess oxygen component of the imput air flow passes through the reaction and exits as oxygen in the exhaust. The quantities of carbon dioxide and water produced are invariant with excess air. They are present in proportion to the amount of carbon and hydrogen in the waste and the waste composition does not change with increased air flow.

Combustion Air Evaluation

Air enters an incinerator from many sources. Underfire and overfire air supplies are often provided and leakages allow air into a furnace from charging doors, equipment coolant flows, and observation ports. Measurement of airflow into an incinerator is an impractical task with most incinerator systems because of these various and uncertain input flows. There is only one exhaust flow, however, and the characteristics of the exhaust can be measured and can be related to inlet airflow.

Instruments for measuring carbon dioxide and/or oxygen in a hot, dirty gas flow are readily available. By measuring the percent oxygen or carbon dioxide in the exhaust the quantity of excess air entering the system can be determined.

Examining the above equation for combustion of cellulose with 50% excess air, the percent carbon dioxide and oxygen can be determined:

$$\frac{6}{6 + 33.86 + 5 + 3} \cdot 100 = 12.53\%\ CO_2 \text{ by volume} \qquad (9\text{-}4)$$

$$\frac{1.63}{1.63 + 5.85 + 0.56 + 0.59} \cdot 100 = 18.89\%\ CO_2 \text{ by weight} \qquad (9\text{-}5)$$

$$\frac{3}{6 + 33.86 + 5 + 3} \cdot 100 = 6.27\%\ O_2 \text{ by volume} \qquad (9\text{-}6)$$

$$\frac{0.59}{1.63 + 5.85 + 0.56 + 0.59} \cdot 100 = 6.84\%\ O_2 \text{ by weight} \qquad (9\text{-}7)$$

The percentages of carbon dioxide and oxygen in the exhaust gas are often given on a dry basis. The measurement of the dry component of the gas is made by passing the gas through a dryer or a low temperature water bath before it is analyzed. The moisture content of a waste is normally variable and by measuring dry gas flow this variable is eliminated. The percent of carbon dioxide and of oxygen in dry flue gas is calculated as follows:

$$\frac{6}{6 + 33.86 + 3} \cdot 100 = 14.00\%\ CO_2\text{, dry, by volume} \qquad (9\text{-}8)$$

$$\frac{1.63}{1.63 + 5.85 + 0.59} \cdot 100 = 20.20\%\ CO_2\text{, dry, by weight} \qquad (9\text{-}9)$$

$$\frac{3}{6 + 33.86 + 3} \cdot 100 = 7.00\%\ O_2\text{, dry, by volume} \qquad (9\text{-}10)$$

$$\frac{0.59}{1.63 + 5.85 + 0.59} \cdot 100 = 7.31\%\ O_2\text{, dry, by weight} \qquad (9\text{-}11)$$

Performing the same calculations for the case where zero excess air is introduced (stoichiometric combustion), the following list summarizes this data, in percentages:

		0% EA	50% EA
By Volume, Dry:	CO_2	21.01	14.00
	O_2	0.00	7.00
By Weight, Dry:	CO_2	29.37	20.20
	O_2	0.00	7.31
By Volume, Wet:	CO_2	17.88	12.53
	O_2	0.00	6.27
By Weight, Wet:	CO_2	26.72	18.89
	O_2	0.00	6.84

With 0% excess air, all of the oxygen supplied will be used in the reaction and no oxygen will appear in the exhaust. As excess air increases, oxygen in excess of the stoichiometric requirement will be found in the exhaust.

The amount of carbon dioxide produced is a function of the amount of carbon combusted, not the air supply. The total flue gas quantity will increase as the excess air increases. Therefore, with the quantity of carbon dioxide present remaining constant while the total flue gas quantity increases, the percentage of carbon dioxide in the exhaust will decrease as the excess air increases.

Estimating Stack Gas Quantities

The quantity of gas exiting an incinerator stack is a function of excess air as well as feed conditions and exhaust gas treatment system design. When defining an emission it is necessary to establish the excess air quantity at which the emission is to be evaluated. If an incinerator particulate emission standard is stated as 0.05 grains per dry standard cubic foot flue gas, for instance, at 50% excess air, this is a much less stringent requirement than the same allowance at 0% excess air (stoichiometric). At 50% excess air the gas flow rate is significantly higher than at stoichimetric conditions and a 0.05 grain loading will, therefore, result in a much greater allowable particulate discharge.

Equation 9-2 represents the combustion of cellulose at stoichiometric conditions. Emissions are normally stated in terms of dry standard cubic feet of exiting flue gas. One cubic foot of air at dry, standard conditions weighs 0.075 pounds. From equation 9-2 combustion of 1 pound of cellulose releases a total of 5.55 pounds of dry gas (1.63 lb CO_2 plus 3.92 lb N_2) or 74 dry standard cubic feet (5.55 lb/0.075 lb per cu ft = 74.) of flue gas at standard conditions. If an emission of 0.05 grains particulate per dry standard cubic foot is allowed (7000 grains = 1 lb) the allowable emission is 0.529 lb per hour per 1000 pounds cellulose per hour ($74 \cdot 0.05 \cdot 1000 / 7000 = 0.529$).

Examining equation 9-3, which describes the combustion of cellulose with 50% excess air, 8.07 pounds of dry gas (1.63 lb CO_2 plus 5.85 lb N_2 plus 0.59 lb O_2 = 8.07) or 107.6 dry standard cubic feet of flue gas at standard conditions is produced per pound of cellulose. For the same emission rate, 0.05 grains per dry standard cubic foot, 0.769 pounds particulate emissions per hour is allowed per 1000 pounds of cellulose.

Tabular Values

As noted above, the effort expended to try to achieve "accurate" calculations in waste incineration quickly reaches a point of no return, where variations in waste quantity and quality negate close analytical procedures. In light of this, waste quality can often be approximated to the necessary degree of accuracy by use of historical data. Composition of a typical sample of solid waste is listed in Table 9–2. The chemical analysis of packaging wastes, which comprise the majority of municipal solid waste constituents, is listed in Table 9–3.

A convenient means of estimating the quantity of flue gas generated from refuse is by use of Table 9–4. The heat release of the waste can be calculated (normally in the range of 4500 to 7000 Btu/lb as received) and with this figure, multiplied by a value from Table 9–4, an estimate of flue gas flow can be made. Likewise, the values in Table 9–5 can also be used to determine expected discharge values.

Table 9–2 Typical Composition of Municipal Solid Waste

Component	Weight Percent
Paper, Other Than Newspaper	25.
Newspaper	14.
Garbage	12.
Yard Wastes	10.
Glass, Ceramic, Stone	10.
Metal	8.
Cardboard	7.
Wood	7.
Textiles	3.
Plastic Film	2.
Leather, Rubber, Molded Plastics	2.
Total	100.%

Source: L. Pryde, *Environmental Chemistry*, 1st ed. (Menlo Park, CA), 346.

Table 9–3 Analysis of Packaging Wastes, Dry Basis, %

Waste	Carbon	Hydrogen	Oxygen	Nitrogen	Sulfur	Inerts*	Btu/lb
Corrugated Boxes	43.73	5.70	44.93	0.09	0.21	5.34	7,429.
Brown Paper	44.90	6.08	47.84	0.00	0.11	1.07	7,706.
Paper Food Cartons	44.74	6.10	41.92	0.15	0.16	6.93	7,730.
Waxed Milk Cartons	59.18	9.25	30.13	0.12	0.10	1.22	11,732.
Plastic Coated Paper	45.30	6.17	45.50	0.18	0.08	2.77	7,703
Newspaper	49.14	6.10	43.03	0.05	0.16	1.52	8,480.
Polyethylene	85.60	14.40	–	–	–	–	19,950.
Vinyl	47.10	5.90	18.60	(chlorine 28.40%)		–	8,730.
Plastic Film	67.21	9.72	15.82	0.46	0.07	6.72	13,846.
Textiles	46.19	6.41	41.85	2.18	0.20	3.17	8,036.
Softwood (pine)	52.55	6.08	40.90	0.25	0.10	0.12	9,150.
Hardwood (oak)	49.49	6.62	43.39	0.25	0.10	0.15	8,682.
Glass bottles	0.52	0.07	0.36	0.03	0.00	99.02	84.**
Metal Cans	4.54	0.63	4.28	0.05	0.01	90.49	742.**

*Ash, glass, metal.

**Btu in labels, coatings, and remains of contents of containers.

Source: B. Baum and C. Parker, *Solid Waste Disposal*. Vol. 1, (Ann Arbor, MI: Ann Arbor Science 1973), p. 43.

Table 9–4 Gas Generated from Refuse, Dry Standard Cubic Feet Per 10,000 Btu

Refuse Component	0% EA*	50% EA	12% CO_2
Newspaper	92	139	156
Brown Paper	90	135	156
Trade Magazine	93	140	161
Corrugated Paper Boxes	91	137	158
Plastic Coated Paper	92	139	158
Waxed Milk Cartons	94	144	135
Paper Food Cartons	93	140	155
Junk Mail	91	138	159
Vegetable Food Waste	100	151	159
Citrus Rinds and Seeds	94	142	161
Meat Scraps, Cooked	92	141	129
Fried Fats	90	137	119
Leather Shoe	96	145	144
Rubber	95	145	130
Vacuum Cleaner Catch	95	144	142
Evergreen Trimmings	92	139	149
Balsam Spruce	95	143	150
Flower Garden Plants	97	146	156
Lawn Grass, Gree	92	139	149
Ripe Tree Leaves	101	152	158
Arithmetic Mean	94	142	149
Standard Deviation	2.9	4.4	11.9

*EA = Excess air

Source: A. Licata, "Impact of Test Measurements on Test Results," In *Proceedings* of the 1976 ASME Conference on Present Status and Research Needs in Energy Recovery from Wastes.

Incinerator Flue Gas Calculations

Using a gross heat balance around an incinerator a relatively accurate determination of flue gas quantities can be made, as follows:

To determine the products of combustion from the burning of 250 tons per day of municipal solid waste with 75% excess air:

Moisture content	25%
Incombustible solids (ash)	10%
Heating value	6,500 Btu/lb as received

Table 9–5 Typical Products of Incineration of Municipal Refuse (Without Controls)

Stack Gases	Lb/ton Refuse	Fraction by Volume (dry)
Carbon Dioxide	1,738.	6.05%
Sulfur Dioxide	1.	22 ppm
Hydrogen Chloride	–	5–500 ppm
Carbon Monoxide	10.	0.06%
Oxygen	2,980.	14.32%
Nitrogen Oxides	3.	93 ppm
Nitrogen	14,557.	79.57%
Total Dry Gas	19,289	100%
Water Vapor	1,400.	
Total Gas	20,689.	
Solids, Dry Basis:		
Grate Residue	471.	
Fly Ash	20.	
Total	21,180	

Source: E. Kaiser, *Refuse Reduction Processes,* U.S. Public Health Service Publication No. 1729 (Washington, DC: U.S. Government Printing Office), p. 93.

The actual input to the furnace, on an hourly basis, will be determined (MBH = million Btu per hour):

Refuse, 250 ton/day	=	20,833 lb/hr
less moisture (25%)	=	5,208 lb/hr
less ash (10%)	=	2.083 lb/hr
Combustibles (moisture/ash free)		13,452 lb/hr
Heating value	=	6,500 Btu/lb
Heat release (6,500 Btu/lb · 20,833 lb/hr)	=	135 MBH

The flue gas generated by combustion, at stoichiometric conditions (using the approximation of 7.5 lb dry flue gas and 0.51 lb moisture produced per 10,000 Btu of heat release):

7.5 lb dry gas/10 KB · 135 MBH = 101,250 lb/hr
0.51 lb H_2O/10 KB · 135 MBH = 6,885 lb/hr

The air required is calculated as follows:

Dry gas from combustion	101,250 lb/hr
Moisture from combustion	6,855 lb/hr
Subtotal	108,135 lb/hr
less combustible input	13,452 lb/hr
Stoichiometric airflow	94,683 lb/hr
plus 75% excess air	71,012 lb/hr
Total	165,695 lb/hr

The flow exiting the incinerator is composed of a dry gas component and a moisture component:

Dry gas from combustion	101,250 lb/hr
plus excess air	71,012 lb/hr
Total dry gas	172,262 lb/hr
Moisture from combustion	6,885 lb/hr
plus moisture from MSW	5,208 lb/hr
Total moisture flow	12,093 lb/hr

To determine the flue gas volumetric flow (CFM) the temperature within the furnace and at the furnace exit must be determined. Table 9–6 lists the enthalpy of air and moisture and Table 9–7 lists the specific volume of air and moisture as a function of temperature. The enthalpy and volume of dry flue gas can be approximated by that of air. For 172,262 lb/hr dry gas and 12,093 lb/hr moisture, taking the datum to be 60°F, and with 3% radiation loss, the temperature at which the enthalpy of the gas flow is equal to the heat contained within the gas (135 MBH less 3% radiation loss equals 131 MBH in the gas flow) is calculated as follows:

Assume a gas temperature of 2,300°F. The enthalpy of the gas flow, per Table 9–6, is:

Dry gas	= 172,262 lb/hr · 596.45 Btu/lb	=	103 MBH
H_2O	= 12,093 lb/hr · 2252.60 Btu/lb	=	27 MBH
	Total	=	130 MBH

Evaluating the flow at 2,400°F:

Dry gas	= 172,262 lb/hr · 625.52 Btu/lb	=	108 MBH
H_2O	= 12,093 lb/hr · 2315.32 Btu/lb	=	28 MBH
	Total	=	136 MBH

Table 9–6 Enthalpy, Air and Moisture

Relative to 60°F			Relative to 80°F	
H air Btu/lb	$H\ H_2O$ Btu/lb	Temp °F	H air Btu/lb	$H\ H_2O$ Btu/lb
21.61	1091.92	150	16.82	1071.91
33.65	1116.62	200	28.86	1096.61
45.71	1140.72	250	40.92	1120.71
57.81	1164.52	300	53.02	1144.51
69.98	1188.22	350	65.19	1168.21
82.19	1211.82	400	77.40	1191.81
94.45	1235.47	450	89.66	1215.46
106.79	1259.27	500	102.00	1239.21
119.21	1283.07	550	114.42	1263.06
131.69	1307.12	600	126.90	1287.11
144.25	1331.27	650	139.46	1311.26
156.87	1355.72	700	152.08	1335.71
169.59	1380.27	750	164.80	1360.26
187.38	1405.02	800	177.59	1385.01
195.26	1430.02	850	190.47	1410.01
208.21	1455.32	900	203.42	1435.31
221.35	1480.72	950	216.46	1460.71
234.36	1506.42	1000	229.57	1486.41
247.55	1532.40	1050	242.76	1512.40
260.81	1558.32	1100	256.02	1538.31
274.15	1584.80	1150	264.36	1564.80
287.55	1611.22	1200	282.76	1591.21
301.02	1638.26	1250	296.23	1618.20
314.56	1665.12	1300	309.77	1645.11
328.17	1692.15	1350	323.38	1672.15
341.85	1719.82	1400	337.06	1699.81
355.58	1747.70	1450	350.82	1727.70
369.37	1775.52	1500	364.58	1755.51
397.17	1832.12	1600	392.33	1812.11
425.08	1890.11	1700	420.29	1870.10
453.24	1948.02	1800	448.45	1928.01
481.57	2007.17	1900	476.78	1987.70
510.07	2067.42	2000	505.28	2047.41
538.72	2128.70	2100	533.93	2108.70
567.52	2189.92	2200	562.73	2169.91
596.45	2252.60	2300	591.66	2232.60
625.52	2315.32	2400	620.73	2295.31
654.70	2377.80	2500	649.91	2357.80
684.01	2443.30	2600	679.22	2423.30
713.42	2511.88	2700	708.63	2491.80

Table 9–7 Specific Volume

T °F	Air ft^3/lb	H_2O ft^3/lb	T °F	Air ft^3/lb	H_2O ft^3/lb
70	13.3	21.5	2000	61.9	99.7
100	14.1	22.7	2100	64.5	103.8
200	16.6	26.8	2200	67.0	107.0
300	19.1	30.8	2300	69.5	111.9
400	21.7	34.9	2400	72.0	116.0
500	24.2	38.9	2500	74.5	120.0
600	26.7	43.0	2600	77.0	124.1
700	29.2	47.0	2700	79.6	128.1
800	31.7	51.1	2800	82.1	132.2
900	34.2	55.1	2900	84.6	136.2
1000	36.8	59.2	3000	87.1	140.3
1100	39.3	63.3	3100	89.6	144.3
1200	41.3	67.3	3200	92.2	148.4
1300	44.3	71.4	3300	94.7	152.5
1400	46.8	75.4	3400	97.2	156.5
1500	49.4	79.5	3500	99.7	160.6
1600	51.9	83.5	3600	102.2	164.6
1700	54.4	87.6	3700	104.7	168.7
1800	56.9	91.6	3800	107.3	172.7
1900	59.4	95.7	3900	109.8	176.8

The temperature corresponding to the heat content of 131 MBH in the gas flow is between 2300°F and 2400°F. By interpolation, therefore:

$$\text{Gas temperature} = 2300°\text{F} + \frac{(131 - 130)}{(136 - 130)}(2400 - 2300) = 2317°\text{F}$$

The flue gas volume can be calculated using Table 9–6:

At 2300°F:

$$\begin{aligned} \text{Dry gas} &= 69.5\ \text{ft}^3/\text{lb} \cdot 172{,}262\ \text{lb/hr} / 60 = 199{,}537\ \text{CFM} \\ H_2O &= 111.9\ \text{ft}^3/\text{lb} \cdot 12{,}093\ \text{lb/hr} / 60 = \underline{22{,}553\ \text{CFM}} \\ \text{Total} &= 222{,}090\ \text{CFM} \end{aligned}$$

After passing through a waste heat boiler the gas would be reduced in temperature to, say, 400°F. The flow at this temperature is:

$$\begin{aligned} \text{Dry gas} &= 21.7\ \text{ft}^3/\text{lb} \cdot 172{,}262\ \text{lb/hr} / 60 = 62{,}301\ \text{CFM} \\ H_2O &= 34.9\ \text{ft}^3/\text{lb} \cdot 12{,}093\ \text{lb/hr} / 60 = \underline{7{,}034\ \text{CFM}} \\ \text{Total} &= 69{,}335\ \text{CFM} \end{aligned}$$

To determine the exiting gas flow in dry standard cubic feet per minute corrected to 50% excess air:

From above, stoichiometric air flow	=	94,683 lb/hr
50% excess air	=	47,341 lb/hr
dry gas from combustion	=	101,250 lb/hr
Total	=	148,591 lb/hr

At standard conditions dry gas (air) weighs 0.075 lb/ft^3:

148591 lb/hr / (0.075 lb/ft^3 · 60) = 33,020 SCFM, 50% XAir

This technique can be applied to determine the flue gas flow from any combustion process.

References and Bibliography

1. Brunner, C. Program solves airstream energy balances. *Chemical Engineering* (November 16, 1981).
2. American Society of Mechanical Engineers. 1974. *Combustion Fundamentals for Waste Incineration*. New York: ASME.
3. National Bureau of Standards. 1972. *Thermodynamic Data for Industrial Incinerators*. Washington, DC: Government Printing Office.
4. Brunner, C. 1984. *Incineration systems: Selection and design*. New York: Von Nostrand Reinhold.
5. Brunner, C. 1980. *Design of sewage sludge incineration system*. Park Ridge, NJ: Noyes.
6. American Gas Association. 1973. *Guidebook for industrial and commercial gas fired incineration*. Arlington, VA: AGA.
7. Keenan, J., and J. Kaye. 1957. *Gas tables*. New York: Wiley.
8. Keenan, J., and F. Keyes. 1957. *Thermodynamic properties of steam*. New York: Wiley.
9. Weber, J. Predict thermal conductivity of pure gases. *Chemical Engineering* (January 12, 1981).
10. American Gas Association. 1973. *Fundamentals of Gas Combustion*. Arlington, VA; AGA, January.
11. Brunner, C. 1982. *Incineration Systems*. Rockville, MD: Government Institutes.
12. Brunner, C., and S. Schwartz. 1983. *Energy and resource recovery from waste*. Park Ridge, NJ: Noyes.
13. Sisson, W. Nomogram estimates combustion air. *Power Engineering* (August 1982).
14. Stewart, E. Determine excess air from wet stack gas analysis. *Pollution Engineering* (April 1979).

10

Estimating Incinerator Emissions

Incinerator emissions control is a function of the nature of the waste charged as well as the incinerator type and the operating parameters of the incinerator. The physical and chemical characteristics of the waste control, the nature of the flue gas discharge, and the control system must suit this discharge.

A waste with a significant plastics component will necessarily generate hydrogen chloride gas in the exhaust. If a wet scrubber is used, hydrochloric acid will be produced in the scrubber water. This potential for acid production and its attendant corrosive action to control system materials may preclude the use of a wet scrubber for emissions control. Likewise, a waste with a relatively high sulfur content is subject to similar considerations, with the potential for producing sulfuric acid in the exhaust.

A hazardous waste requires high temperatures for effective destruction, as described in an earlier chapter. A baghouse installation is sensitive to temperatures in excess of 300°F to 600°F and therefore cannot be used where relatively high temperatures exit the furnace system.

Gaseous wastes normally produce negligible particulate matter when incinerated: the burning of liquid waste will usually generate less particulate in the exhaust gas stream than the burning of solid waste. For many fluid (liquid and/or gaseous) wastes temperatures of destruction can be extremely high, as compared with solids, and the presence of high temperatures encourage the formation of nitrogen oxides and other gaseous pollutants. Control system selection must address removal of gases, or of particulate as primary concerns.

Physical characteristics of a waste material such as mean size and size variability are normally accommodated by the type of incinerator chosen for its destruction. The chemical characteristics of a waste are generally more crucial to selection of an emission control system than are its physical characteristics.

Incinerator Type

The type of incinerator selected for waste destruction is perhaps the most important factor in estimating emissions generation. The following listing is arranged from the most severe to least severe emissions discharge from a furnace to the control system.

- Pyrolysis: The products of combustion of a pyrolytic reactor are, by their very nature, rich in unburned hydrocarbons, including organic acids and heavy and light organic components. The gas is also heavily laden with particulate matter.
- Fluid Bed Incinerator: All of the ash component of a waste and elutriated sand leaves the furnace through the exhaust when burning the majority of wastes disposed of in this furnace system.
- Rotary Kiln: The action of the kiln is to tumble the waste charge throughout the kiln length. This turbulence results in a relatively high airborne particulate loading.
- Multiple Hearth Furnace: Sludge waste is rabbled, or churned, across a flat hearth, then drops to a lower hearth as gas rushes up to wash the falling material. This action can result in as much as 30% of the ash load becoming airborne, entering the exhaust gas stream.
- Teepee Burner: Solid waste burns practically uncontrolled. The burning process is incomplete and significant quantities of unburned hydrocarbons, carbon monoxide, and particulate are released in the off-gas.
- Single Chamber Incinerator: Unless the temperature within the incinerator is sufficiently high, the waste will normally be improperly combusted, releasing particulate matter and unburned organics to the exhaust gas stream. Even with high temperatures, the exhaust gas will require emissions control.
- Multiple Chamber Incinerator: Sufficient impediments to the exhaust gas flow path are normally provided to drop out the majority of entrained particulate from the gas stream. The turbulence thus produced tends to increase the burnout of unburned organics within the off-gas.
- Central Disposal Incinerators: These incinerators normally operate at high chamber temperatures which tend to destroy organics with a relatively high efficiency. Depending on waste introduction (grate, pneumatic, RDF, etc.) particulate carryover may be high. Where boiler tubes are in the gas stream, a significant percentage of the larger sized particulate matter will be collected from the off-gas and drop in the boiler hopper(s).
- Controlled Air Incinerator: There is minimal agitation of waste and a secondary chamber is provided for controlled burnout of the products of combustion.

Tabular Values

The burning process generally involves combustion of carbonaceous materials in the presence of air, which contains nitrogen as well as oxygen. Analytical methods can be used for forcasting CO, NO, and NO_2 emissions, which will always be present in combustion processes. These methods are discussed later in this chapter. The determination of particulate and other gaseous emissions is much more difficult, as it is

Table 10–1 Estimated Rates of Emission of Contaminants From Fossil Fuels (pounds of contaminants per ton of fuel burned)

Contaminant	Coal (12,000 Btu/lb)	Oil (19,000 Btu/lb)	Gas (1,000 Btu/ft^3)
Particulates	30–150	28	–
Sulfur Oxides (as SO_2)	60–120	60–120	–
Nitrogen Oxides (as NO_2)	8	8–36	14
Acids (as Acetic)	30	27	2.6
Aldehydes (as Formaldehyde)	–	2.6	2.0
Other Organics (+Hydrocarbons)	20	9.2	2.8

Source: American Industrial Hygiene Association, *Air Pollution Manual*, Part 1 (Evaluation), 2d Ed. (Westmont, NJ; AIHA, 1972), 21.

Table 10–2 Estimated Rates of Emission of Contaminants From Internal Combustion Engines (pounds of contaminants per ton/10mb of fuel burned)

Contaminant	Gasoline (1)	Gasoline (2)	Diesel (1)	Diesel (2)
Particulate	0.1	.02	34	9
Sulfur Oxides (As SO_2)	2–5	.5–1.2	10	2.6
Nitrogen Oxides (As NO_2)	25	6	49	13
Acids (As Acetic)	0.6	0.14	10	2.6
Aldehydes (As Formaldehyde)	5.6	1.33	5	1.35
Other Organics (+Hydrocarbons)	151	34	n/a	–

Note: 1. Pound of emissions per ton of fuel.
2. Pound of emissions per 10 million Btu (10MB), the approximate heating value of 1 ton of refuse.

Source: American Industrial Hygiene Association, *Air Pollution Manual*, Part 1 (Evaluation), 2d ed. (Westmont, NJ: AIHA, 1972), 21.

generally waste specific. For instance, SO_2 emissions will not be generated from every combustion process. The potential for sulfur emissions is dependent upon the presence and amount of sulfur compounds in the waste streams.

Forcasting emissions from incineration requires the examination of historical data; for example, evaluation of data from existing incinerators and sources. As a basis for evaluating data from the burning of refuse, Table 10–1 lists contaminants from the burning of fossil fuels, related to ton of fuel and heating value of fuel. Table 10–2 indicates the generation rates of contaminants from internal combustion engines (automobiles and trucks).

The American Industrial Hygiene Association has estimated the generation rate of contaminants from the burning of refuse. Table 10–3

Table 10–3 Emission Factors Without Controls (pounds per ton of refuse burned)

Part One

Pollutant	Municipal Multiple Chamber	Industrial and Commercial Single Chamber	Industrial and Commercial Multiple Chamber
Aldehydes	1.1	5–64	0.3
Benzo(a)pyrene	6,000μg/ton	100,000 μg/ton	500,000μg/ton
Carbon Monoxide	0.7	20–200	0.5
Hydrocarbons	1.4	20–50	0.3
Nitrogen Oxides	2.1	1.6	2.0
Sulfur Oxides	1.9	n/a	1.8
Ammonia	0.3	n/a	n/a
Organic Acids	0.6	n/a	n/a
Particulate	6–12*	20–25	4.0

n/a = not available
*–6 with, 12 without spray chamber

Source: *Air Pollution Manual.* American Industrial Hygiene Association, Part 1 (Evaluation), 2d ed. (Westmont, NJ: AIHA, 1972), 29.

Table 10–4 Emission Factors Without Controls (pounds per ton of refuse burned)

Part Two

Pollutant	Flue Fed Apartment Incinerator	Domestic Single Chamber Without Aux Fuel	Domestic Single Chamber With Aux Fuel
Aldehydes	5.0	6.0	2.0
Benzo(a)pyrene	n/a	n/a	n/a
Carbon Monoxide	n/a	300.	n/a
Hydrocarbons	40.0	100.	1.5
Nitrogen Oxides	0.1	1.5	2.0
Sulfur Oxides	0.5	2.0	2.0
Ammonia	0.4	0.4	negligible
Organic Acids	22.0	13.	4.0
Particulate	26.0	39.	6.0

n/a = not available

Source: American Industrial Hygiene Association, *Air Pollution Manual*, Part 1 (Evaluation), 2d ed. (Westmont, NJ: AIHA, 1972), 29.

presents this data for the burning of refuse in municipal multiple chamber incinerators, industrial and commercial single and multiple chamber incinerators. Contaminants from domestic incinerators burning refuse are listed in Table 10–4. Emission factors for open burning are listed in Table 10–5.

Table 10–5 Emission Factors Without Controls (pounds per ton of refuse burned)

Part Three

Pollutant	Burning Dump	Backyard Burning
Aldehydes	4.0	3.6
Benzo(a)pyrene	25,000μg/ton	350,000μg/ton
Carbon Monoxide	n/a	n/a
Hydrocarbons	280.	280.
Nitrogen Oxides	0.6	0.5
Sulfur Oxides	1.2	0.8
Ammonia	2.3	1.6
Organic Acids	1.5	1.5
Particulate	47.	150.

n/a = not available

Source: American Industrial Hygiene Association, *Air Pollution Manual,* Part 1 (Evaluation), 2d. ed. (Westmont, NJ: AIHA, 1972), 30.

A set of data collected by the U.S. Environmental Protection Agency (EPA) is listed in Table 10–6 which relates generation of particulate and other contaminants to the burning of refuse in various types of incinerators. Estimates of emissions from a refuse burning incinerator downstream of a control device are listed in Table 10–7.

Calculating Carbon Monoxide Generation

Carbon monoxide (CO) is produced where insufficient oxygen is provided to completely combust a fuel or other combustible material. It is therefore indicative of burning, or combustion efficiency. The greater the amount of air present, and the greater the degree of turbulence within the burning chamber, the less carbon monoxide will be formed.

Turbulence as a combustion parameter cannot be easily quantified. The amount of air present and the combustion temperature affect the equilibrium constant, from which the relationship of CO and CO_2 produced for a given reaction can be found. Table 10–8 is derived from equilibrium analyses and lists the formation of CO as a function of excess air, temperature, and the ratio of carbon to hydrogen in the fuel.

Note that the generation of CO decreases with an increase in the air supply, and increases with an increase in temperature. As expected, with increased carbon in the fuel there will be an increase in the rate of formation of CO.

Table 10–6 Emission Factors For Refuse Incinerators Without Controls (pounds per ton as charged)

Incinerator Type	Particulate	SO_x	CO	Hydrocarbon	NO_x
Municipal:					
Multiple Chamber,					
Uncontrolled	30	2.5	35	1.5	3
With Settling Chamber and Sprays	14	2.5	35	1.5	3
Industrial/Commercial:					
Multiple Chamber	7	2.5	10	3	3
Single Chamber	15	2.5	20	15	2
Controlled Air	1.4	1.5	neg	neg	10
Flue Fed:					
Single Chamber	30	0.5	20	15	3
With Afterburner	6	0.5	10	3	10
Domestic Single Chamber:					
Without Second. Burner	35	0.5	300	100	1
With Secondary Burner	7	0.5	neg	2	2
Pathological	8	neg	neg	neg	3

neg—negligible

Source: U.S. Environmental Protection Agency, *Compilation of Air Pollutant Emission Factors*, 3rd ed. *U.S. EPA PB-223996* (Washington, DC: Government Printing Office, April 1979), 2.1–3.

Burning methane, CH_4, the quantity of carbon monoxide formed can be estimated as follows:

$$\begin{array}{ccccc} 16.05 & 32.00 & 44.01 & 18.02 & 28.01 \\ CH_4 & +\ 2\ O_2 \rightarrow & CO_2 & +\ 2\ H_2O\ + & CO\ (\text{trace}) \\ 16.05 & 64.00 & 44.01 & 36.04 & \\ 1.00 & 3.99 & 2.74 & 2.25 & \end{array}$$

The equilibrium equation defines the stoichiometric oxygen required for burning methane, 3.99 per pound of methane. (The numbers above the equilibrium equation represent the atomic weights of the individual molecules. The first line beneath the equation is the total molecular weight and beneath that is the total molecular weight normalized to 1 pound of fuel, methane in this case.)

Assuming that the above reaction will take place at 2500°F, with 20% excess air supplied, the amount of CO formed per pound of stoichiometric air is 1.972E-5, from Table 10–8. For one pound of methane, therefore:

Table 10–7 Emission Factors For Refuse Burning Downstream of an Electrostatic Precipitator, Typical Values

Pollutant	Concentration PPM[a]	Concentration gr/dscf[b]	Pounds/ Ton
Particulate		0.02	0.34
Sulfur Dioxide	80		2.4
Nitrogen Oxides as NO_2	75		1.6
Carbon Monoxide	150		1.9
Hydrocarbons	16		0.12
Hydrochloric Acid	200		3.4
Fluorides	6.5		0.06
Lead		0.00068	0.012
Mercury		0.00024	0.0064
Beryllium		0.000000003	0.000000051
Sulfuric Acid Mist		0.0023	0.04
Tetrachlorodibenzo-p-dioxins			0.00000001
Polynuclear Aromatics			0.00001
Polychlorinated Biphenyls			0.00013
Asbestos		—not detected—	
Hydrogen Sulfide		—not detected—	
Vinyl Chloride		—not detected—	
Reduced Sulfur		—not detected—	

[a]parts volume per million parts volume
[b]corrected to 12% CO_2

Source: W. O'Connell, G. Stotler, and R. Clark, *Emissions and Emission Control in Modern Municipal Incinerators,* In: *Proceedings* of the 1982 National Waste Processing Conference, ASME.

$$3.99 \text{ lb } O_2 \text{ required}$$

$$3.99 \text{ lb } O_2 \cdot 4.3197 \cdot \frac{\text{lb Air}}{\text{lb } O_2} = 17.24 \text{ lb Air}$$

$$17.24 \text{ lb Air} \cdot 1.972 \cdot 10^{-5} \cdot \frac{\text{lb CO}}{\text{lb Air}} = 0.00034 \text{ lb CO}$$

Therefore, 0.00034 pound CO is theoretically produced when burning 1 pound of methane at 2500°F with 20% excess air.

To determine the theoretical volume of CO produced in parts per million parts (ppm) of methane, the volumes of each component must be calculated. Using the perfect gas law:

$$W = \frac{144 \cdot P}{R \cdot T}$$

Table 10–8 Generation of Carbon Monoxide (CO) Pounds CO Per Pound Stoichiometric Air

	Excess Air							
Temp. °F	0	10%	20%	30%	50%	100%	150%	200%
For C_1H_0 (Carbon)								
100	—	—	—	—	—	—	—	—
1500	2.060E-07	1.407E-09	1.039E-09	8.827E-10	7.344E-10	5.997E-10	5.474E-10	5.193E-10
1832	9.544E-06	1.480E-07	1.093E-07	9.287E-08	7.725E-08	6.303E-08	5.758E-08	5.474E-08
2192	1.170E-04	6.110E-06	4.506E-06	3.827E-06	3.183E-06	2.597E-06	2.370E-06	2.249E-06
2500	4.830E-04	5.128E-05	3.778E-05	3.207E-05	2.666E-05	2.176E-05	1.985E-05	1.883E-05
3000	3.418E-03	9.796E-04	7.248E-04	6.159E-04	5.118E-04	4.174E-04	3.807E-04	3.610E-04
For C_3H_4 ($C_1H_{1.33}$)								
1000	—	—	—	—	—	—	—	—
1500	2.591E-08	1.080E-09	7.968E-10	6.755E-10	5.606E-10	4.554E-10	4.149E-10	3.928E-10
1832	8.131E-06	1.137E-07	8.375E-08	7.105E-08	5.896E-08	4.791E-08	4.362E-08	4.133E-08
2192	8.858E-05	4.692E-06	3.454E-06	2.930E-06	2.429E-06	1.975E-06	1.796E-06	1.702E-06
2500	3.870E-04	5.468E-05	2.897E-05	2.456E-05	2.036E-05	1.653E-05	1.505E-05	1.425E-05
3000	2.792E-03	7.567E-04	5.574E-04	4.723E-04	3.912E-04	3.175E-04	2.889E-04	2.735E-04
For CH_4								
1000	2.919E-10	—	—	—	—	—	—	—
1500	8.122E-08	7.970E-10	5.418E-10	4.589E-10	3.799E-10	3.075E-10	2.794E-10	2.642E-10
1832	5.572E-06	7.749E-08	5.699E-08	4.828E-08	3.997E-08	3.236E-10	2.940E-08	2.778E-07
2192	6.161E-05	3.199E-06	2.350E-06	1.991E-06	1.647E-06	1.333E-06	1.204E-06	1.144E-06
2500	2.743E-04	2.686E-05	1.972E-05	1.669E-05	1.380E-05	1.116E-05	1.014E-05	9.581E-06
3000	2.022E-03	5.174E-04	3.801E-04	3.213E-04	2.654E-04	2.146E-04	1.946E-04	1.839E-04

Note: 15E-03 = 0.0015

where W is in lb/ft^3 = m/V (mass per unit volume)
and M is weight, in pounds
and V is volume, cubic feet
with R = (1545 m)/M where M is the molecular weight

Therefore:

$$W = \frac{1545\, m}{144 \cdot P \cdot M}$$

For the case in question, the volume of CO versus the volume of CH_4, both gases are at the same temperature and pressure. Therefore the volume ration is simplified to the following:

$$\frac{V_{CO}}{V_{CH_4}} = \frac{m_{CO}}{m_{CH_4}} \cdot \frac{M_{CH_4}}{M_{CO}}$$

$$= \frac{0.00034 \text{ lb}}{1.0 \text{ lb}} \cdot \frac{16.05}{28.01} \cdot 1{,}000{,}000$$

$$= 195 \text{ ppm by volume, CO to } CH_4.$$

To relate the CO present to the total flue gas flow, at 20% excess air, the flue gas generation must be determined. Noting that 20% excess air corresponds to 1.2 · 2 = 24 molecules of oxygen, which carries with it 2.4 · 3.7619 = 9.03 molecules of nitrogen from the air.

16.05	32.00	28.02		44.01	18.02	32.00	28.02
CH_4 +	$2\,O_2$ +	$0.03\,N_2$	→	CO_2 +	$2\,H_2O$ +	$0.4\,O_2$ +	$9.03N_2$
16.05	7680	253.02		44.01	36.04	12.80	253.02
1.00	4.79	15.76		2.74	2.25	0.80	15.76

The outlet flue gas volume is:

$$V_{FG} = \frac{1545}{144} \cdot \frac{T}{P} \cdot \left[\frac{m_{CO_2}}{M_{CO_2}} + \frac{m_{H_2O}}{M_{H_2}} + \frac{m_{O_2}}{M_{O_2}} + \frac{m_{N_2}}{M_{N_2}}\right]$$

The ratio of CO to flue gas volume is therefore:

$$\frac{V_{CO}}{V_{FG}} = \frac{m_{CO}/M_{CO}}{m_{FG}/M_{FG}}$$

$$= \frac{1{,}000{,}000 \cdot 0.00034 \text{ lb}/28.01 \text{ lb}}{2.74/24.01 + 2.25/18.02 + 0.80/16.00 + 15.76/14.01}$$

$$= 8.91 \text{ ppm, by volume, CO in the exiting gas flow.}$$

Calculating Generation of Nitrogen Oxides

The amount of nitrogen oxides generated is a function of temperature, excess air, and fuel composition. The greater the amount of air, the greater will be the amount of nitrogen present, and higher quantities of NO_x

will be expected. The series of graphs in Figure 10–1 show this relationship. The scale pound NO_x per million Btu relates to fuel heat release and is directly proportional to the stoichiometric oxygen demand, which is the second vertical scale.

Tables 10–9 and 10–10 list generation rates for NO and NO_2 respectively. To illustrate the use of these charts, from the previous example for the burning of methane, it was found that 17.24 pound of air was the stoichiometric demand for burning 1 pound of methane. Methane is CH_4, which has a carbon component between that of C_3H_4 and C_0H_4 listed in Tables 10–9 and 10–10. A closer examination of these tables, however, indicates that NO and NO_2 production is not significantly

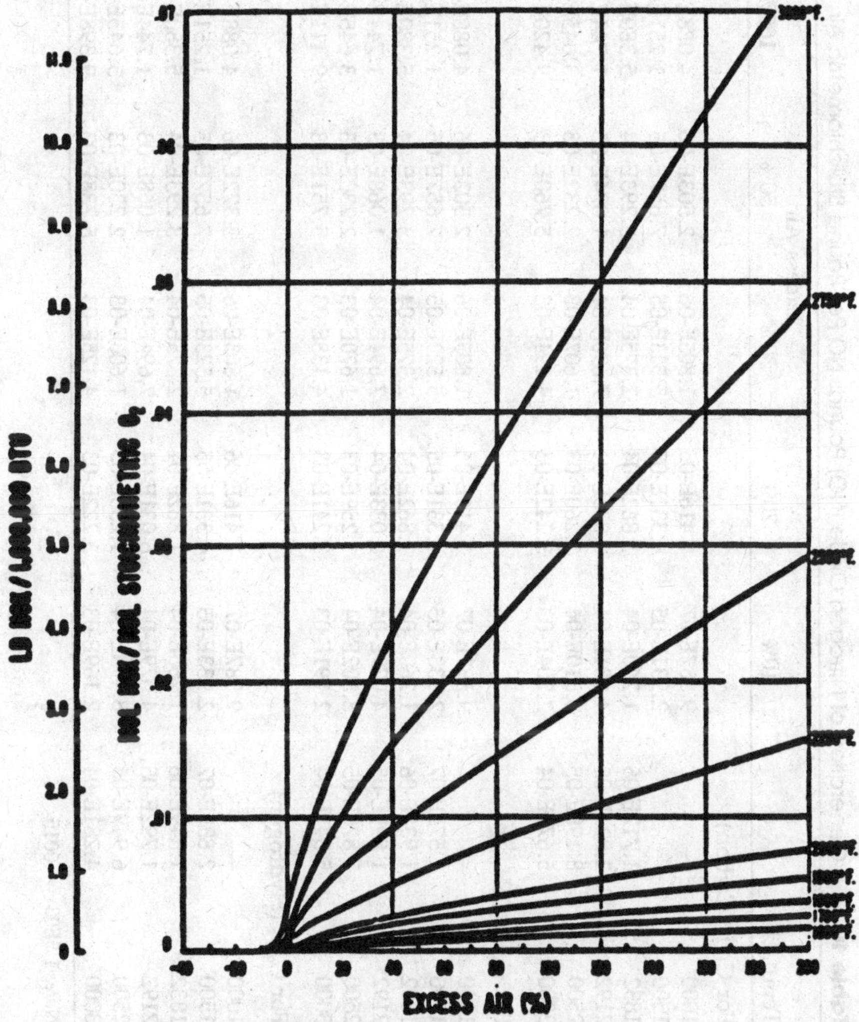

Figure 10–1 Generation of NO_x.

Table 10–9 Generation of Nitrogen Oxide (NO) Pounds NO Per Pound Stoichiometric Air

	Excess Air							
Temp. °F	0	10%	20%	30%	50%	100%	150%	200%
For C_1H_0 (Carbon)								
1000	—	9.587E-07	1.416E-06	1.805E-06	2.503E-06	4.088E-06	5.599E-06	7.082E-06
1500	—	2.931E-05	4.331E-05	5.523E-05	7.657E-05	1.251E-04	1.712E-04	2.166E-04
1832	1.712E-06	1.260E-04	1.862E-04	2.375E-04	3.293E-04	5.380E-04	7.366E-04	9.316E-04
2192	2.022E-05	4.081E-04	6.033E-04	7.696E-04	1.074E-03	1.744E-03	2.388E-03	3.022E-03
2500	8.198E-05	8.510E-04	1.260E-03	1.607E-03	2.232E-03	3.645E-03	4.993E-03	6.318E-03
3000	5.677E-04	2.204E-03	3.242E-03	4.194E-03	5.760E-03	9.420E-03	1.291E-02	1.633E-02
For C_3H_4 ($C_1H_{1.33}$)								
1000	—	9.587E-07	1.416E-06	1.805E-06	2.503E-06	4.088E-06	5.599E-06	7.082E-06
1500	2.575E-07	2.933E-05	4.331E-05	5.523E-05	7.657E-05	1.251E-04	1.712E-04	2.166E-04
1832	1.932E-06	1.260E-04	1.862E-04	2.375E-04	3.293E-04	5.380E-04	7.366E-04	9.316E-04
2192	1.899E-05	4.079E-04	6.033E-04	7.694E-04	1.068E-03	1.744E-03	2.388E-03	3.022E-03
2500	7.870E-05	8.502E-04	1.259E-03	1.670E-03	2.230E-03	3.645E-03	4.993E-03	6.318E-03
3000	5.341E-04	2.191E-03	3.241E-03	4.138E-03	5.751E-03	9.411E-03	1.290E-02	1.632E-02
For C_0H_2 (Hydrogen)								
1000	—	9.587E-07	1.416E-06	1.805E-06	3.372E-06	4.088E-06	5.599E-06	7.082E-06
1500	2.835E-07	2.933E-05	4.331E-05	5.523E-05	7.657E-05	1.251E-04	1.712E-04	2.166E-04
1832	1.849E-06	1.260E-04	1.862E-04	2.375E-04	3.293E-04	5.380E-04	7.366E-04	9.316E-04
2192	1.792E-05	4.079E-04	6.031E-04	7.694E-04	1.068E-03	1.744E-03	2.388E-03	3.018E-03
2500	6.949E-05	8.493E-04	1.258E-03	1.606E-03	2.230E-03	3.645E-03	4.993E-03	6.315E-03
3000	4.281E-04	2.169E-03	3.222E-03	4.125E-03	5.738E-03	9.398E-03	1.288E-02	1.632E-02

Note: 1.5E03 = .0015

Table 10–10 Generation of Nitrogen Dioxide (NO_2) Pounds NO_2 Per Pound Stoichiometric Air

	Excess Air							
Temp. °F	0	10%	20%	30%	50%	100%	150%	200%
For C_1H_0 (Carbon)								
1000	—	1.208E-07	2.416E-07	3.622E-07	6.039E-07	1.208E-06	1.812E-06	2.416E-06
1500	—	4.045E-07	8.086E-07	1.213E-06	2.022E-06	4.045E-06	6.065E-06	8.089E-06
1832	—	6.528E-07	1.352E-06	2.029E-06	3.386E-06	6.768E-06	1.015E-05	1.354E-05
2192	6.538E-10	1.018E-06	2.008E-06	3.066E-06	5.113E-06	1.024E-05	1.536E-05	2.049E-05
2500	1.375E-08	1.330E-06	2.671E-06	4.015E-06	6.701E-06	1.343E-05	2.015E-05	2.668E-05
3000	1.359E-07	1.863E-06	3.718E-06	5.589E-06	9.341E-06	1.875E-05	2.816E-05	3.758E-05
For C_3H_4 ($C_1H_{1.33}$)								
1000	—	1.180E-07	2.364E-07	3.552E-07	5.936E-07	1.192E-06	1.793E-06	2.395E-06
1500	—	3.948E-07	7.916E-07	1.189E-06	1.988E-06	3.991E-06	6.006E-06	8.020E-06
1832	—	6.598E-07	1.324E-06	1.989E-06	3.326E-06	6.681E-06	1.005E-05	1.342E-05
2192	6.961E-10	9.944E-07	1.998E-06	3.004E-06	5.027E-06	1.010E-05	1.262E-05	2.031E-05
2500	1.219E-08	1.296E-06	2.611E-06	3.932E-06	6.585E-06	1.325E-05	1.994E-05	2.664E-05
3000	1.172E-07	1.798E-06	3.615E-06	5.453E-06	9.151E-06	1.846E-05	2.782E-05	3.718E-05
For C_0H_2 (Hydrogen)								
1000	—	1.107E-07	2.228E-07	3.362E-07	5.656E-07	1.149E-06	1.740E-06	2.335E-06
1500	—	3.705E-07	7.460E-07	1.126E-06	1.894E-06	3.848E-06	5.826E-06	7.820E-06
1832	—	6.189E-07	1.247E-06	1.883E-06	3.168E-06	6.438E-06	9.751E-06	1.309E-05
2192	—	9.318E-07	1.882E-06	2.843E-06	4.787E-06	9.734E-06	1.475E-05	1.980E-05
2500	1.471E-08	1.213E-06	2.457E-06	3.718E-06	6.269E-06	1.276E-05	1.934E-05	2.597E-05
3000	7.021E-08	1.654E-06	3.376E-06	5.123E-06	8.679E-06	1.775E-05	2.694E-05	3.655E-05

Note: 1.5E-03 = .0015

affected by fuel composition at the 20% excess air and 2500°F parameters used in this example. Therefore, for NO use 1.258E-03 and for NO_2 use 2.5E-06 pounds generated per pound of stoichiometric air.

For NO (atomic weight of 30.01):

$$17.24 \text{ lb Air} \cdot 1.258 \cdot \frac{10^{-3} \text{ lb NO}}{\text{lb Air}} = 0.022 \text{ lb NO}$$

$$\frac{V_{NO}}{V_{CH_4}} = \frac{m_{NO}}{m_{CH_4}} \cdot \frac{M_{CH_4}}{M_{NO}} \cdot 1{,}000{,}000 = \frac{0.022 \text{ lb}}{1.0 \text{ lb}} \cdot \frac{16.05}{30.01} \cdot 1{,}000{,}000$$

$$= 11{,}766 \text{ ppm by volume, NO to } CH_4$$

For NO_2 (atomic weight of 46.01):

$$17.24 \text{ lb Air} \cdot 2.5 \cdot \frac{10^{-6} \text{ lb } NO_2}{\text{lb Air}} = 0.0000431 \text{ lb } NO_2$$

$$\frac{V_{NO_2}}{V_{CH^4}} = \frac{m_{NO^2}}{m_{CH_4}} \cdot \frac{M_{CH^4}}{M_{NO_2}} \cdot 1{,}000{,}000 = \frac{}{1.0 \text{ lb}} \cdot \frac{0.022 \text{ lb}}{46.01} \overset{16.05}{\cdot} 1{,}000{,}000$$

$$= 15 \text{ ppm by volume, } NO_2 \text{ to } CH_4$$

To obtain the total weight of NO_x expressed as NO_2:

NO: 0.022 lb NO · 46.01/30.01	=	0.034 lb NO_2 equivalent
NO_2:	=	0.0000431 lb
Total NO_x		0.034 lb per lb CH_4

Similarly, the volumetric presence of NO_x is:

11,766 ppm NO · 46.01/30.01	=	18,034 ppm NO_2 equivalent
NO_2	=	15 ppm NO_2
Total NO_x		18,049 ppm NO_x by volume

To relate the NO, NO_2, and NO_x generated to the exiting gas flow use the methods in the previous discussion of CO generation.

Chromium Release

Chromium is usually generated in trivalent or hexavalent form from a burning process. Hexavalent chromium, as discussed in chapter 4, has been found to be hazardous, whereas the trivalent form is not. The formation of hexavalent chromium increases as the oxygen (air) content in the furnace increases. If the combustion air decreases to the point where insufficient air is available for complete destruction of the organics present (starved-air condition) trivalent chromium will predominate.

The majority of chromium, generated as an oxide, will remain in the residual (in the ash) however some of this residual will necessarily be airborne, and will be discharged from a stack.

References and Bibliography

1. Rollins, R., and J. Homolya. Measurement of gaseous hydrogen chloride emissions from municipal refuse energy recovery systems in the United States. *Environmental Science and Technology*, (November 1979) 13 11: 1380–1383.
2. Law, S., and G. Gordon. Sources of metals in municipal incinerator emissions. *Environmental Science and Technology*, (April 1979) 13/4: 432–438.
3. Suprenant, N. 1981. *Emissions assessment of conventional stationary combustion systems. EPA 600/S7-81-003/C.* Washington, DC: Government Printing Office, August.
4. Wall, H. 1981. Thermal conversion of municipal wastewater sludge, Phase II, Study of heavy metal emissions. *EPA 600/2-81-203*, Washington, DC: Government Printing Office, September.
5. Gerstle, R., and D. Albrinck. Atmospheric emissions of metals from sewage sludge incineration. *APCA Journal* (November 1982) 32/11: 1119–1123.
6. Bennett, R., and K. Knapp. Characterization of particulate emissions from municipal wastewater sludge incinerators. *Environmental Science & Technology* (December 1982).
7. Higher sludge incineration temperatures may increase metal emissions. *Sludge Newsletter* (September 26, 1982): 189.
8. Air emissions: Coal, oil, incinerators. *Solid Wastes Management* (January 1983).
9. USEPA. 1979. Measurement of PCB emissions from combustion sources. *EPA 600/7-79-047*, Washington, DC: Government Printing Office, February.
10. Tow, P. A. suggested change in method for determining emissions from incinerators. *APCA Journal* (December 1980), 30/12: 1326–1327.
11. USEPA. 1979. Test methods to determine the mercury emissions from sludge incineration plants. *EPA 60/4-79-058*. Washington, DC: Government Printing Office, September.
12. Rinaldi, G. 1979. *An Evaluation of Emission Factors for Waste to Energy Systems.* USEPA. Washington, DC: Government Printing Office.
13. Jahnke, J. A research study of gaseous emissions from a municipal incinerator. *APCA Journal* (August 1977).
14. USEPA. 1973. *Compilation of air pollutant emission factors. EPA AP42.* Washington, DC: Government Printing Office, April.
15. Greenberg, R., Zoller, W., and Gordon, G. Atmospheric emissions of elements from the parkway sewage sludge incinerator. *Environmental Science and Technology* (August 1981) 15/1: 64–70.

16. Copeland, B. A study of heavy metal emissions from fluidized bed incinerators. Presented at the Purdue Industrial Waste Conference, Princeton, NJ, May 1975.
17. Parks, D. and E. Fletcher, Formation and emission of nitric oxide in fluidized bed combustion. *Environmental Science and Technology* (August 1975).
18. Stevens, J. Energy recovery and emissions from municipal waste incineration. Ontario APCA, September 1981.
19. Shen, T. Air pollutants from sewage sludge incineration. *ASCE Journal* (February 1979): 64–74.
20. Brunner, C. 1984. *Incineration systems: Selection and design*. New York: Van Nostrand Reinhold.

11

Gas Cleaning

The public perception of an incinerator is generally negative. People expect smoke and odor, and before the passage of the Clean Air Act in the past decade and its implementation, the public concern was indeed justified. It is incumbent upon the system designer to comply with the existing air pollution control statutes to control nuisance emissions from incinerators and to keep in mind that even an occasional failure of these systems will be met with severe public reaction.

There are numerous types and sizes of air pollution control equipment on the market today. They range from the unsophisticated, a series of baffles, to the relatively complex high-energy water scrubbing devices, utilizing alkali, to clean gas streams of solid, liquid, and gaseous pollutants.

Removal of pollutants is not necessarily for the sake of regulations alone. In many cases incinerator discharges must be free of nonregulated components in order to safeguard downstream equipment from excessive wear from erosion (physical wear of materials) and/or corrosion (chemical degradation of materials).

The generation of pollutants is a function of the following factors:

- Waste Composition
- Charging Rate
- Method of Charge
- Furnace Type
- Furnace Design
- Burning Conditions (Three Ts of Temperature, Turbulence, Time)
- Excess Air Introduced into the Incinerator

The more common pollutants that are discharged from incinerators and which must be substantially removed from the gas stream are the following:

- Particulate Matter
- Chlorides
- Sulfur Oxides

Other pollutants are not normally present in incinerator discharges to the extent that they warrant particular attention. If they are present in any significant amounts, however, they will often be removed by the systems for control of the three pollutants identified above.

Particle Size

Any particle control device will have its efficiency vary as a function of particle catch size. Particulate matter is normally measured by mean physical size in units of one millionth of a meter, that is a micron (μ).

Particle size generated from an incineration process cannot be calculated. It can be measured and estimates can only be inferred from prior data based on measurement. As would be expected, therefore, data on particle size for a particular application is rare. It is a function of waste properties, furnace design and furnace operation.

Figure 11–1 identifies the full range of particle sizes, from fractions of a micron to a thousand microns in diameter. The so-called diameter is

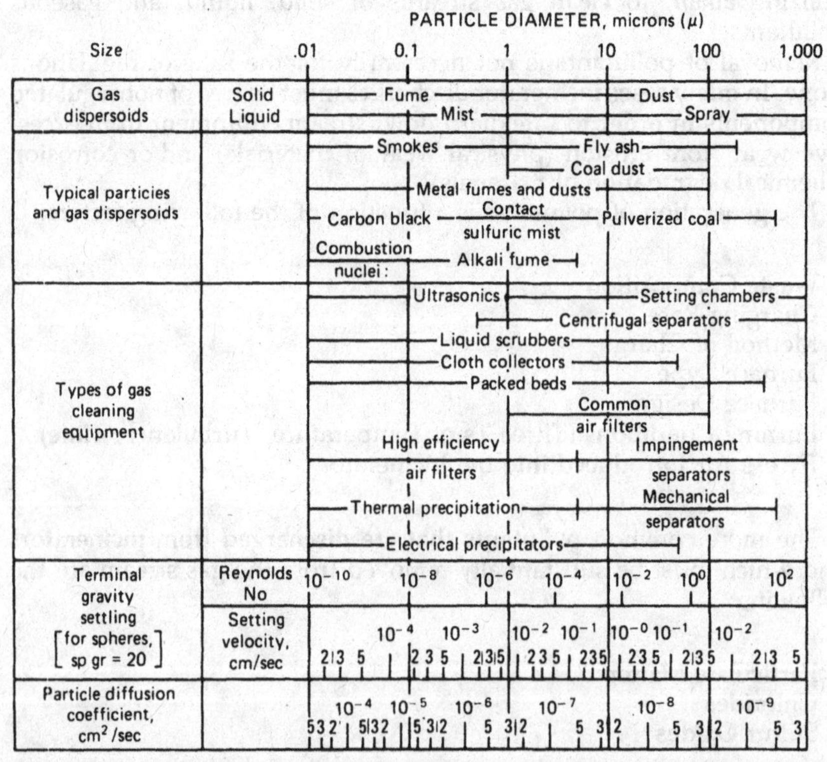

Figure 11–1 Particulate matter by size. (Source: M. First and P. Dreiner, "Concentrations of particulates found in air." *Industrial Hygiene and Occupational Medicine* 5 (1952): 387.

not truly a diameter because particles are not truly round. The diameter referred to is the mean diameter; that is, it can be thought of as the diameter of an opening which is just large enough to pass the particle in question.

Fumes and mists, as shown in Figure 11–1, are generally defined as solid or liquid matter respectively, with mean particle size below 1μ. Dust and sprays are likewise defined as in excess of 1μ mean diameter. There are a number of types of air cleaning devices that are shown to be effective for smaller particle size removal such as the electrostatic precipitator (ESP), cloth collectors, and liquid scrubbers. The "common air filters" referred to are those found on commercial and domestic air circulating or ventilating systems.

Removal Efficiency

The efficiency of a particulate cleaning device is simply the weight of material removed divided by the weight of that quantity of particulate matter entering the device. It can be calculated for a specific size of particulate or can be based on total particulate loading.

A typical calculation of removal efficiency can be made using Table 16–3, as follows:

Total flow into control device: 217.01 lb/hr
Particles less than 1.1μ:

$$0.9\% \cdot 217.01 \text{ lb/hr} = 1.95 \text{ lb/hr}$$

Total flow out of control device: 1.48 lb/hr
Particles less than 1.1μ:

$$67.7\% \cdot 1.48 \text{ lb/hr} = 1.00 \text{ lb/hr}$$

Removal efficiency of particles less than 1.1μ:

$$\frac{1.95 - 1.00}{1.95} \cdot 100 = 48.72\%$$

Removal efficiency of total particulate load:

$$\frac{217.01 - 1.48}{217.01} \cdot 100 = 99.32\%$$

As would be expected, the efficiency of a scrubber decreases with decreased particle size. Less than 50% of the smaller particles are removed while the overall removal efficiency for all particulate matter exceeds 99%.

The efficiency of specific types of gas cleaning equipment for various particulate size distributions is discussed in the appropriate chapter describing that equipment.

Acid Removal

Industrial waste will often have a halogen or sulfide component. Municipal wastes have relatively high proportions of plastics and many of these materials have a significant of chlorides. Chlorides and/or sulfides in an incinerator charge will necessarily produce hydrogen chloride and sulfur trioxide gas in the off-gas stream. These two gases will readily dissolve in water to form hydrochloric acid and sulfuric acid respectively. Both of these acids will attack and corrode steel and many other materials. Hazardous waste incineration regulations require that acid gas neutralization systems be required when significant quantities of hydrogen chloride is present in a exhaust (in excess of 4 lb/hr HCl). There are no other federal regulations specifically addressing the emissions of acidic components from incinerators. The control of these emissions is necessary for the protection of downstream equipment.

The mechanism of removal includes solubility plus chemical reactivity. Table 11–1 lists solutions commonly used for the removal of acidic and other gases and gaseous components from the gas stream. Note that HCl and sulfur dioxide are both soluble in water. Water alone will absorb these compounds. Adding sodium hydroxide to water will increase absorbtion of HCl, and to illustrate this process note the following equation:

$$\begin{array}{cccc} HCl & + NaOH \rightarrow & H_2O & + NaCl \\ 36.46 & 40.00 & 18.42 & 58.44 \\ 1.00 & 1.09 & 0.49 & 1.60 \end{array}$$

Hydrogen chloride (HCl), will combine with sodium hydroxide (NaOH), to produce water and sodium chloride. As is evident from this equation, the hydrogen chloride in the gas stream, on the left side of the equation, combines with liquid sodium hydroxide, which is normally dissolved in an aqueous (water) solution. The water in the system plus that produced by this reaction, containing soluble sodium chloride, also produced by this reaction, in effect has cleansed the gas stream of its acidic, chloride component.

The first line beneath the above equation is the atomic weight of each compound and the second line is the atomic weight normalized to 1 pound of HCl (divide the first line by 36.46, the atomic weight of hydrogen chloride, to obtain the second line). Therefore, the theoretical weight of sodium hydroxide required to neutralize 1 pound of hydrogen chloride is apparent; that is, 1.09 pounds. Likewise, 1.60 pounds of sodium chloride is produced per 1 pound of hydrogen chloride.

As noted, the NaOH required for this reaction is a theoretical figure. The actual neutralization process is a function of many factors such as the following:

Table 11–1 Gas Absorption Systems of Commercial Importance

Solute	Solvent	Reagent	Degree of Commercial Importance High	Moderate	Low
CO_2, H_2S	Water	none	x		
CO_2, H_2S	Water	Monoethanolamine	x		
CO_2, H_2S	Water	Diethanolamine	x		
CO_2, H_2S	Water	Triethanolamine			x
CO_2, H_2S	Water	Diaminoisopropanol			x
CO_2, H_2S	Water	Methyl diethanolamine			x
CO_2, H_2S	Water	K_2CO_3, Na_2CO_3	x		
CO_2, H_2S	Water	NH_3		x	
CO_2, H_2S	Water	NaOH, KOH		x	
CO_2, H_2S	Water	K_3PO_4		x	
HCl, HF	Water	none	x		
HCl, HF	Water	NaOH	x		
Cl_2	Water	none	x		
SO_2	Water	none			x
SO_2	Water	NH_3		x	
SO_2	Water	Xylidine		x	
SO_2	Water	Dimethylaniline		x	
SO_2	Water	$Ca(OH)_2$, oxygen			x
SO_2	Water	Aluminum hydroxide		x	
SO_2	Water	Aluminum sulfate		x	
NH_3	Water	none	x		
HCN	Water	NaOH	x		
CO	Water	Copper ammonium salts	x		

Source: C. Brunner, *Incineration Systems: Selection and Design* (New York: Van Nostrand Reinhold, 1984).

- Concentration of HCl in the Gas Stream
- Temperature of the Gas Stream
- Concentration of NaOH in Water
- Presence of Other Compounds in the Water Flow
- Water Temperature
- Contact Time of Gas and Water Flow

An acid neutralization system should be designed with a safety factor of at least two regarding the ratio of reagent to solute (NaOH to HCl in the above example) in order to assure satisfactory mixing and effective contact time for removal. It is important to note that many acid neutralization systems will absorb carbon dioxide as well as acid gas. Incinerator off-gas streams will necessarily contain a high volume of CO_2 and this should be considered in the design of a neutralization system. Standard chemical engineering texts should be consulted to determine solution equilibrium constants, obtaining parameters of design which would not result in wasting a neutralizing agent in the absorbtion of CO_2 rather than efficient acid removal.

12

Inertial Systems

Air discharges from incineration systems contain particulate matter, gases, odor, and noise. Inertial systems provide a relatively inexpensive means to remove larger particulate matter from the gas stream. They can be used in conjunction with other control equipment and normally provide insufficient reduction in emissions to be the sole removal device. They cannot provide the control that current statutes demand. Dry inertial systems are effective only with particulate matter,and when properly designed they can provide some noise attenuation. Wet systems can be used to reduce gas or odor discharges.

Settling Chambers

A settling chamber is a simple device that is successful in removing larger particulate matter from a gas stream. It is a stilling chamber where the velocity of a gas is decreased or remains unchanged from the previous process. Heavier particulate, generally above 100μ (100-micron mean particle diameter), will tend to settle out of the stream and will

Table 12–1 Settling Velocity in Still Air

Particle Size, μ	In/Hr	Ft/Sec
0.01	Negligible	—
0.10	0.047	—
0.5	1.4	—
1.	5.	—
5.	107.	0.002
10.	426.	0.010
50.	—	0.247
100.	—	0.987
500.	—	9.250
1000.	—	13.167

μ = 1 micron: 1,000,000μ = meter

Note: The above table is based on the fall of a spherical body of specific gravity of unity.

Source: W. Frank, American Air Filter Co., Inc., Louisville, KY, undated.

fall to the floor of the chamber. Table 12–1 lists the settling velocity of various size particles in still air. A 100μ particle will have a settling velocity of approximately 1 foot per second. To obtain a magnitude of the size of a chamber that would collect this particle by settlement, assume the gas is flowing through the chamber at 20 feet per second. If the chamber were 2 feet high a particle flowing beneath the top of the chamber would fall to the bottom in (2 feet)/(1 foot per second) = 2 seconds. At a flow rate of 20 feet per second the chamber would have to be (20 feet/sec)·(2 seconds) = 80 feet in length. A chamber 2 feet high by 40 feet long is impractical, and this is a minimum length. If eddies or other turbulent conditions appear within the gas stream the settling velocity can been significantly higher than that listed.

For much larger particles, 500 μ for instance (the size of raindrops), the chamber that would be required would be only 4 feet 4 inches in length. A chamber 2 feet by 4 feet is a reasonable size for incinerator application.

The above calculations, utilizing Table 12–1, can be used to determine the dropout of particulate along a length of flue or breeching. It is approximate, but the value can be used to estimate an order of magnitude for soot collection and it can also provide information on the frequency of clean-up required for maintenance of the flue system.

A settling chamber is not a true inertial device. Separation of particulate from the gas stream with this equipment is more a function of the particle geometry, that is, physical dimensions, than the particle inertia, or weight.

Baffles

Baffles are walls or plates inserted in the flow stream to act as walls which are used as targets for the flowing particulate matter. Particles within a gas stream, generally above 15μ, will be stopped by a baffle, whereas the gas stream itself will flow around the obstruction. After hitting a baffle the particulate matter will drop to the chamber floor and will collect there for subsequent removal.

Dry baffles are subject to caking of particulate which can create housekeeping problems and which may lead to unwanted burning of the caked layer. To avoid these problems some types of incinerators are provided with wet baffles. Water washes the baffle surface to prevent a particle buildup. Water will also tend to remove more particulate matter from the gas stream by sorbtion, and can also absorb condensible contaminants within the gas.

When a wet baffle system is used there are a number of important factors to consider. The spent water can be laden with particulate matter, organic acids, and other materials. Their presence may make the wastewater unsuitable for discharge in a sewer or other conventional system. A treatment facility is often required to clean the water before it is discharged to a public waterway or sewer. Also, a supply of water must be provided for the baffle. The incinerator design must accommodate the presence of a water supply. If water comes in contact with hot refractory for instance, the refractory could fail or be severely damaged.

Baffles are frequently placed within incinerator chambers or at the exit of an incinerator to reduce the carryover of larger particulate into the gas stream. Where wet baffles are used they are used on small incinerator systems, usually on-site incinerators burning no more than 1500 or 2000 pounds of waste per hour. The use of baffles reduces the particulate loading on downstream equipment, and thus reduces their maintenance requirements, increases their efficiency, and allows smaller downstream equipment sizes to be used.

Dry Cyclonic Separators

The cyclone is a true inertial separator. A typical cyclonic separator is shown in Figure 12–1. Gas entering the unit forms a vortex which eventually reverses and forms a second vortex when leaving the cyclonic chamber. Particulate within the gas stream, because of their inertia relative to the gas stream, tend to move toward the outside wall. They drop from the wall, the inside of the cyclone, to an external receiver for eventual disposal. The particulate collected by a cyclonic collector usually has a high unburned fraction. The larger particles collected are indicative of unburned waste particles airborne within the gas stream. Collected soot is often reinjected into the incinerator where the unburned fraction can be burned to destruction.

Cyclones will remove larger particles (greater than 15μ), but will have negligible effect on smaller particle sizes. A cyclone collector is often placed before another control device such as an electrostatic precipitator

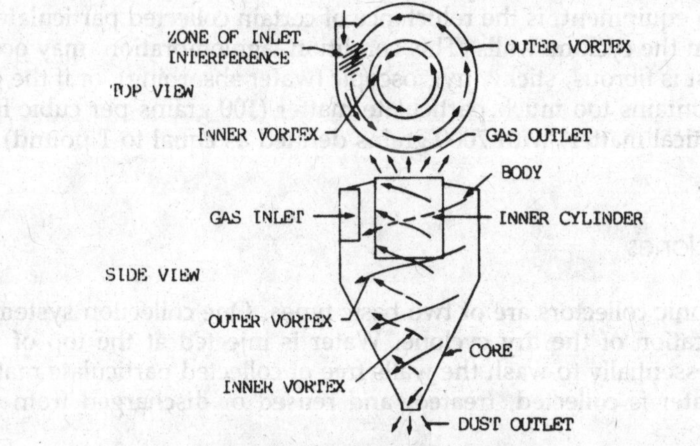

Figure 12–1 Cyclone separator.

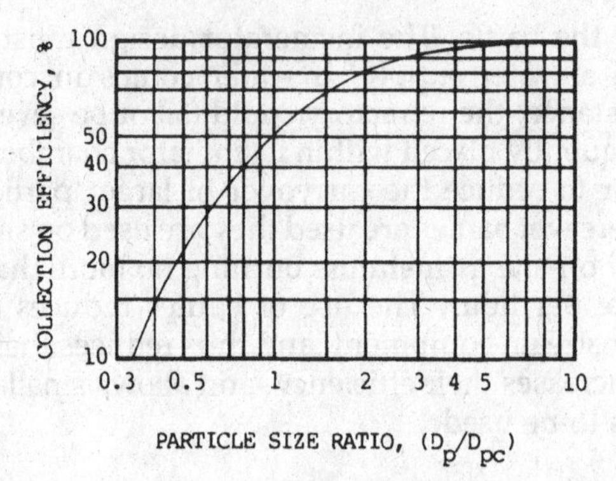

Figure 12–2 Cyclone efficiency vs. particle size ratio.

or a baghouse. The cyclone will remove larger particles from the gas stream resulting, as with the use of baffles, in increased efficiency of the downstream equipment.

Of note is that the larger particles are usually the greatest weight component of particulate in a gas stream. For instance, 85% of the emission weight may be over 15μ with only 15% below 15μ. Use of a cyclonic collector, which has a good removal efficiency for larger particles, will allow provision of less efficient downstream control devices while providing a high overall removal efficiency, based on total particle size loading.

Figure 12–2 shows the relationship between particle size and collection efficiency. With D_{pc} the particle cut size, the ratio $(D_p)/(D_{pc})$ is the actual particle size, D_p, related to the particle cut size. The cut size is the mean diameter of those particles collected by a particular piece of equipment with 50% efficiency. Note that collection efficiency drops off rapidly as the particle size decreases.

One danger in the use of cyclonic separators, as with other dry inertial collection equipment, is the reluctance of certain collected particulate to drop from the cyclone walls. This condition, agglomoration, may occur if the dust is fibrous, sticky, hygroscopic (water absorbing), or if the gas stream contains too much particulate matter (100 grains per cubic foot as a practical matter, with 7000 grains defined as equal to 1 pound).

Wet Cyclones

Wet cyclonic collectors are of two basic types. One collection system is a modification of the dry cyclone. Water is injected at the top of the cyclone essentially to wash the walls free of collected particulate mater. Spent water is collected, treated, and reused or discharged from the system.

A second type of wet cyclonic separator, also termed a low-energy cyclonic scrubber, utilizes a damper at the gas entrance to the cyclone. Water is injected into the gas stream upstream of the cyclone, and upstream of the damper. The damper restricts flow entering the cyclone chamber and imparts a high velocity, or spin, to the gas stream. The entrance to the cyclone is tangential and the result of the "spin damper" at the cyclone entrance is a relatively fast cyclonic flow within the unit. The discharge is tangential, not axial as with the above double-vortex cyclonic separator. Additional water is normally injected within the cyclone for additional cleaning of the gas stream.

Figure 12–3 is a typical cyclonic scrubber with a spin damper. The spin damper is manually controlled to vary the velocity of the gas stream entering the cyclone. Sprays within the cyclone provide additional motion, and mixing, of the gas stream. The water injection process tends to quench the gas as well as clean it and the gas outlet temperature from the cyclone will normally be in the range of 150°F to 180°F.

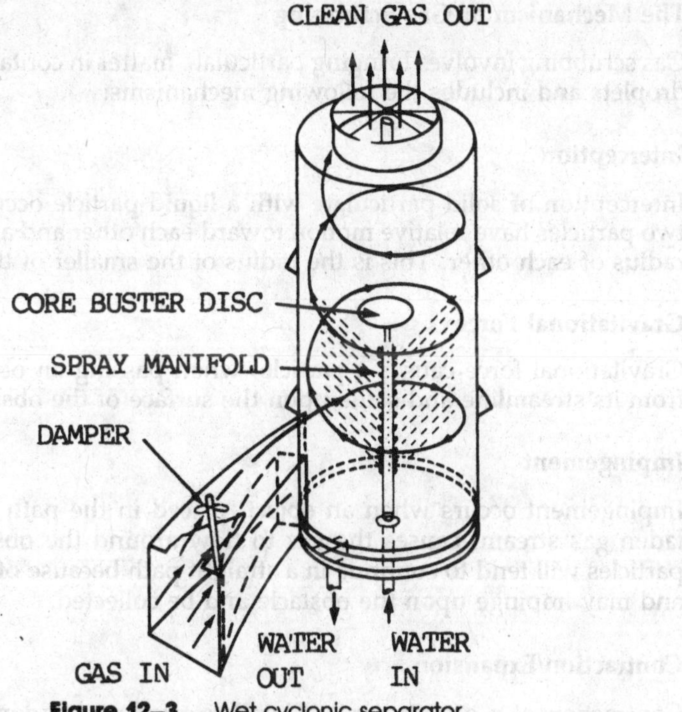

Figure 12–3 Wet cyclonic separator.

13

Wet Gas Scrubbers

Inertial systems take advantage of the weight, or mass, of a particle relative to that of the gas stream carrying it. Wet scrubbing systems, in contrast, utilize a scouring or turbulence to remove particles from a gas stream. A scrubbing action is primarily a function of particle geometry or physical dimension and not mass. An inertial system may be used in conjuction with a scrubber to facilitate the removal of moisture from the gas stream.

The Mechanism of Gas Scrubbing

Gas scrubbing involves bringing particulate matter in contact with liquid droplets and includes the following mechanisms:

Interception

Interception of solid particulate with a liquid particle occurs when the two particles have relative motion toward each other and are within one radius of each other. This is the radius of the smaller of the particles.

Gravitational Force

Gravitational force causes a particle, when passing an obstacle, to fall from its streamline and settle upon the surface of the obstacle.

Impingement

Impingement occurs when an object, placed in the path of a particle-laden gas stream, causes the gas to flow around the obstacle. Larger particles will tend to continue in a straight path because of their inertia, and may impinge upon the obstacle and be collected.

Contraction/Expansion

Contraction of a gas stream will tend to produce condensation of the moisture within the stream. High turbulence within a contracted area will result in good contact between solid particulate and the liquid particles. The dust-laden liquid particles will have the same velocity as the rest of the gas stream and then passing through an area of expansion,

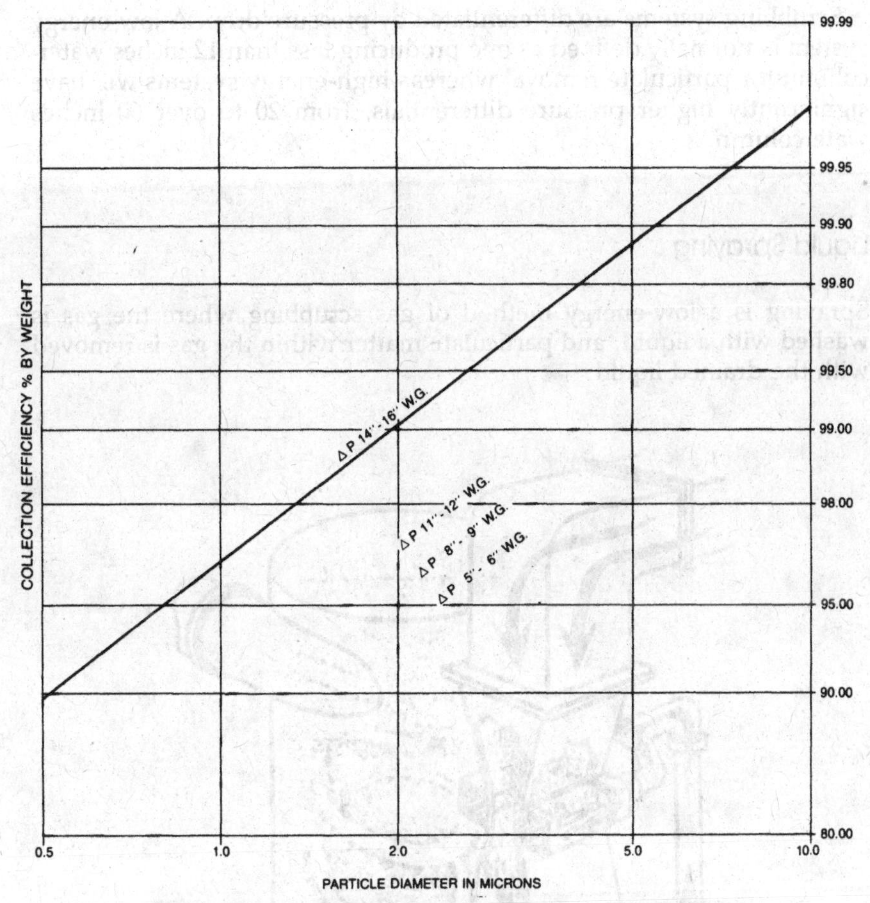

Figure 13–1 Collection efficiency versus particle size.

these particles will continue their direction of flow while the balance of the gas stream can be directed to flow in another direction. In effect, this process of expansion/contraction produces good separation of particulate matter from the gas stream.

The above mechanisms are normally all present at the same time in gas scrubbing to a greater or lesser degree.

The effectiveness of a scrubbing system is usually directly related to the pressure drop across the scrubber. The higher the pressure drop, the greater the turbulence/mixing and, therefore, the more effective the scrubbing action. This feature is illustrated by the graph in Figure 13–1. For a 2μ particle, for instance, a pressure differential of 8 inches (WC) (Water column equivalent to water gauge, WG) will result in a removal efficiency of 95% whereas a 35-inch WC differential will provide almost total (99.9%) removal from the gas stream.

Scrubbing systems are differentiated by pressure drop. A low-energy system is normally defined as one producing less than 12 inches water-column for particulate removal whereas high-energy systems will have significantly higher pressure differentials, from 20 to over 60 inches watercolumn.

Liquid Spraying

Spraying is a low-energy method of gas scrubbing where the gas is washed with a liquid, and particulate matter within the gas is removed with the drained liquid.

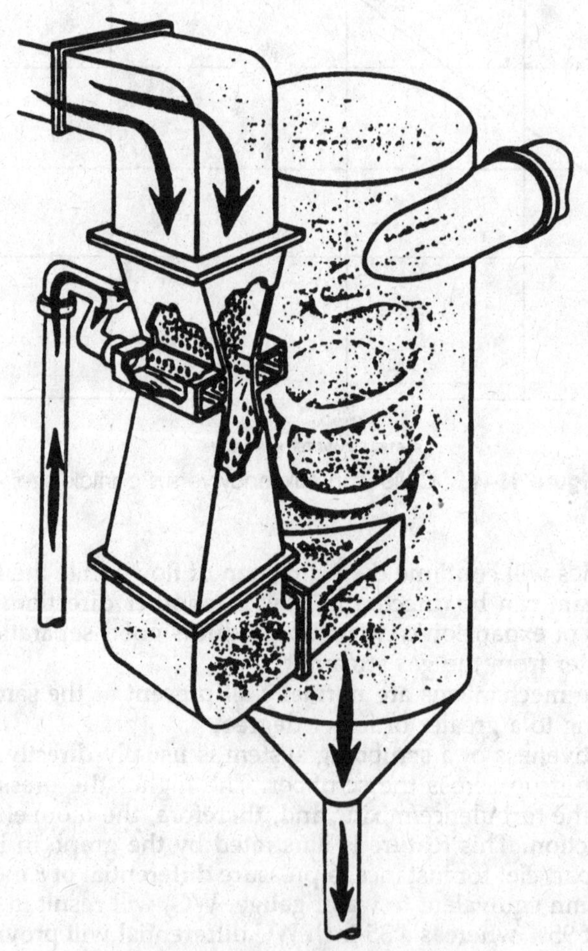

Figure 13–2 Venturi scrubber.

A spray directed along the path of dust or other particulate matter will impinge upon these particles with an efficiency directly proportional to the number of droplets and to the force imparted to the droplets. This results in a droplet size range; that is, the smaller the size of the individual droplete the greater their number. However, the smaller the droplet particle size the less force is associated with them. The optimum water droplet particle size for particulate scrubbing is approximately 100μ. Above 100μ there are too few particles and below this size the droplets have insufficient force.

The mechanism of diffusion promotes deposition of dust particles on water droplets. Diffusion, or Brownian motion, is that property of materials of different diameters to intermingle although the materials may initially be at rest, much as natural gas diffuses within a contained room, although the air within the room is at rest. In scrubbing, diffusion helps the particulate and liquid droplets come in contact and its effectivity is inversely proportional to the size of water and/or solid particulate. The smaller the particle or droplet size, the more rapid the diffusion, that is, the more rapid the wetting process will be.

Spray chambers are at times used as low-cost means of removing heavier particles from the gas stream. Water is sprayed at rates of from 3 to 8 gpm per 1000 cfm of gas flow entering the chamber. The heavier particulate matter is wetted and drops out of the gas stream.

Venturi Scrubber

Venturi scrubbers are widely used where water is readily available as high efficiency, high energy gas cleaning devices. The heart of this system is a venturi throat where gases pass through a contracted area, reaching velocities of 200 to 600 feet per second, and then pass through an expansion section. From the expansion section the gas enters a large chamber for separation of particles or for further scrubbing. Water is injected at the venturi throat or just upstream of the venturi section. Figure 13–2 illustrates a venturi scrubber where water is injected at its throat. For this design from 5 to 7 gallons of water per 1000 cubic feet of gas is normally required.

An adjustable venturi throat is shown in Figure 13–3. Two throat flaps are illustrated; however, it is often designed with a single flap. Water is injected into a "precooler" section immediately before the throat. The precooler quenches the gases to their adiabatic temperature, normally below 200°F, The throat area is adjustable and is normally controlled (manually or automatically) to maintain a desired pressure drop. Note that in this illustration a vane or chevron demister (water separator) is included in the outlet chamber to remove entrained water particles.

A flooded disc scrubber design (also referred to as a plumb-bob actuator) is illustrated in Figure 13–4. The conical plug is positioned to increase the gas flow area, similar to the throat flap design. Typical dimensions

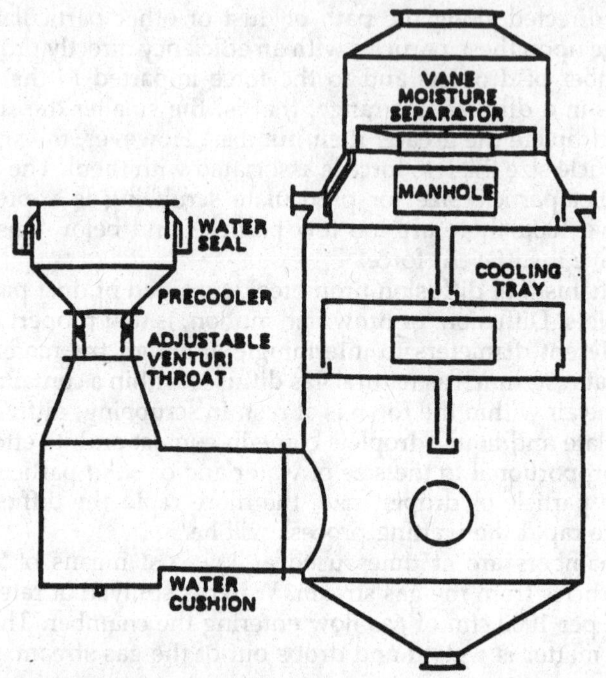

Figure 13–3 Adjustable throat venturi scrubber.

for this unit are given in Table 13–1 for a range of gas flows entering the system. Air enters the scrubber through flange "A" shown in this illustration. Water is injected above the venturi section, tangentially to the inlet gas flow.

A variation of the conventional venturi scrubber is shown in Figure 13–5. Compressed air is injected into the gas stream, with water, and the equipment design is such that the water is immediately atomized. Gas scrubbing with the atomized, turbulent water spray occurs rapidly and efficiently. This design minimizes the use of water although a source of compressed air must be provided. The unit shown contains a venturi section (the "contact cylinder"), an expansion chamber, and a chevron (baffle type) demister.

A serious problem associated with venturi sections is the erosive effect of the gas/liquid mixture passing through the throat. As noted previously, the throat velocity is extremely high creating the potential for wear of the throat surface through erosion. In addition, the corrosive effect of the wet, acidic gas flow is heightened by its turbulent action.

Venturi scrubbers are normally used to remove particles from a gas stream down to approximately 2μ in diameter. It is normally uneconomical to provide the amount of pressure differential required for further removal within a venturi. A tray scrubber can provide greater removal of small particulate (under 5μ) than a venturi for a given pressure drop.

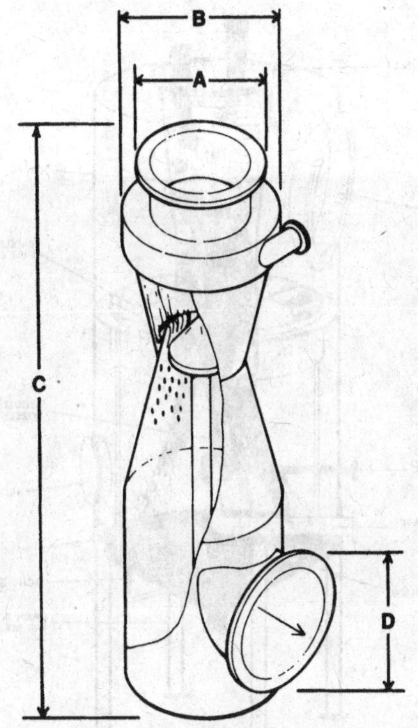

Figure 13–4 Flooded disc scrubber.

Table 13–1 Flooded Disc Scrubber: Typical Dimensions

Gas Volume (ACFM)	A	B	C	D
2,500	1 ft 0 in	3 ft 6 in	6 ft 9 in	1 ft 1 in
5,000	1 ft 4 in	3 ft 10 in	7 ft 11 in	1 ft 6 in
10,000	1 ft 10 in	4 ft 4 in	8 ft 10 in	2 ft 0 in
25,000	3 ft 0 in	5 ft 6 in	10 ft 4 in	3 ft 3 in
50,000	4 ft 2 in	7 ft 2 in	13 ft 7 in	4 ft 5 in
75,000	5 ft 2 in	8 ft 2 in	14 ft 10 in	5 ft 7 in
100,000	5 ft 11 in	8 ft 11 in	16 ft 3 in	6 ft 0 in
150,000	7 ft 3 in	10 ft 3 in	18 ft 3 in	7 ft 6 in
350,000	11 ft 0 in	13 ft 6 in	25 ft 0 in	11 ft 0 in

It must be noted that the greater the pressure drop the higher the fan power requirement. For instance, if a fan draws 100 horsepower when providing the required flow at 20 inches WC differential, it would have to be increased to 250 horsepower if the differential were doubled to 40 inches WC.

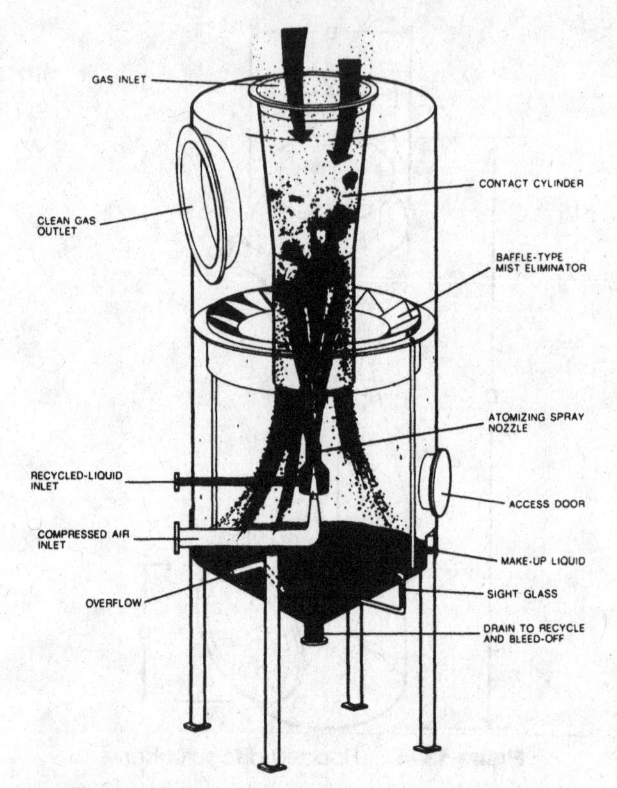

Figure 13–5 Air injection scrubbing device.

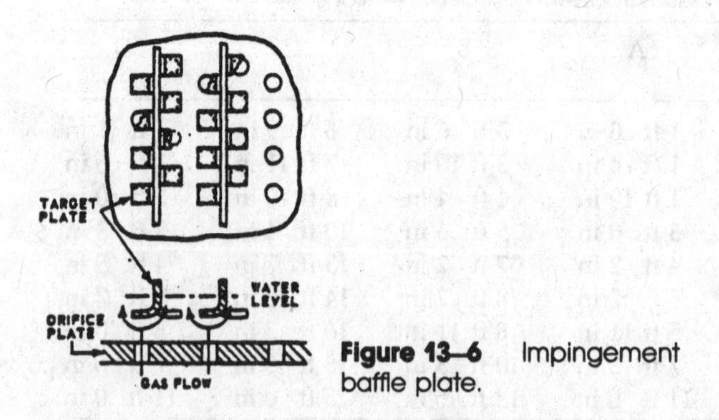

Figure 13–6 Impingement baffle plate.

Tray Scrubbers

Impingement tray scrubbers are essentialy perforated plates with target baffles. Tray scrubbers without impingement plates have no large gas directing baffles but are simly perforated plates within a tower, usually immediately downstream of a venturi. Figure 13–6 illustrates a typical impingement plate. A water level is maintained above the trays (there are usually two or more trays). The geometrical relationship of the tray thickness, hole size, and spacing, and the impinger details, result in a high efficiency device for the removal of small size particulate, less than 5μ. In addition to gas cleaning, these units provide effective gas cooling. The intimate mixing of gas and water create excellent heat transfer between the two.

As many as 300 openings are provided per square foot of tray area. The openings can be from 1/16 inch to 3/8 inch in diameter. Gas flows up through the openings while water is trying to flow, by gravity, counter to the gas flow. Highly effective turbulence with attendant mixing of, or scrubbing of, the solid particulate within the water/water spray environment effectively catches the small micron particulate and removes it from the gas steam.

Tray scrubbers will have a pressure drop of from 2 to 3 inches WC per tray. Impingement trays will add another 1/2 to 1 inch WC differential per tray.

Normally a venturi scrubber or an inertial separator is placed immediately in front of a tray scrubber. Larger particulate is removed by the upstream equipment. If larger particles entered a tray scrubber the tray perforations would quickly clog requiring a significant maintenance (cleaning) effort and reducing equipment availability.

Self-Induced Scrubber

There are a number of scrubbers that employ a unique geometry, utilizing the gas flow to generate scrubbing action. For instance, in the unit shown in Figure 13–7 the water level controls the scrubbing action of the gas. The higher the water level, the greater, or longer, the contact

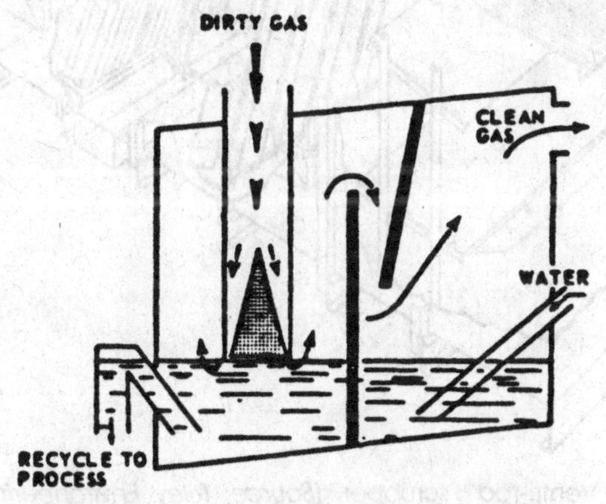

Figure 13–7 Self-induced spray scrubber.

between the gas stream and the water particles. As the water level decreases, however, the gas discharge into the water bath will generate a surface effect, atomizing some of the water surface. The net effect of this system is to obtain relatively good particulate removal efficiency while utilizing relatively low water flows and low gas differential pressures.

Typical differential pressures across this unit are in the range of 4 to 8 inches WC. With this pressure drop removal efficiencies have been found equivalent to that obtained with 14 to 18 inches WC in a conventional venturi section.

Ventri-Rod©

The Ventri-Rod© scrubber is a relatively simple device employing fixed rods placed across the gas stream to produce a scrubbing effect. Figure 13–8 illustrates this sytem. Gas enters the top of the unit, directed through

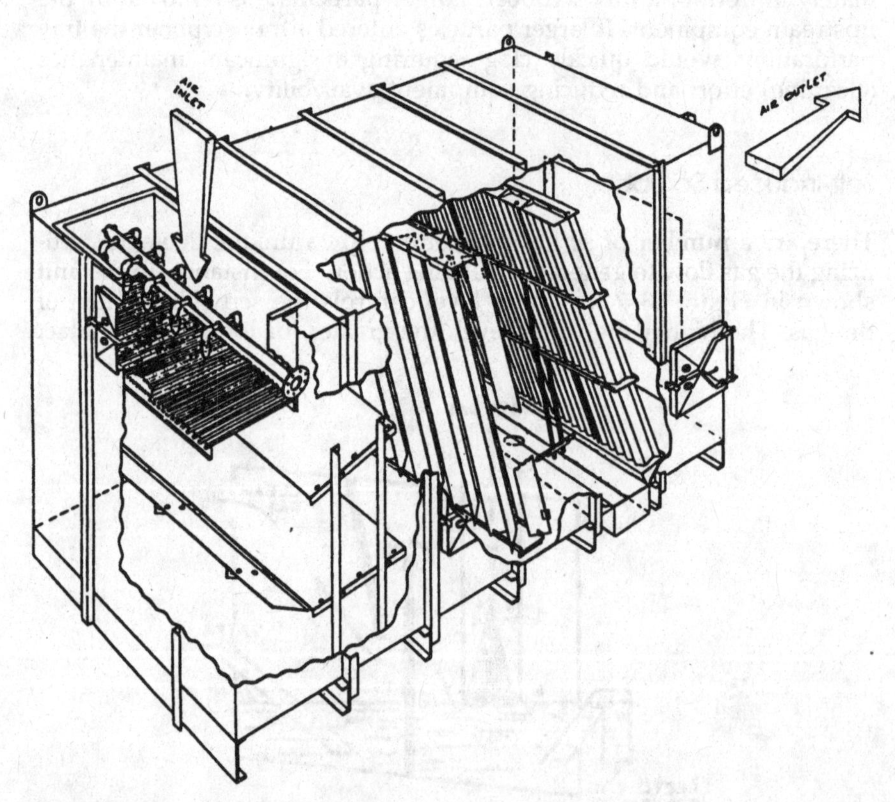

Figure 13–8 Ventri-rod© scrubber. (*Source: Riley Environeering, Inc., Shiller Park, IL.*)

Table 13–2 Ventri-rod Scrubber Typical Dimensions

Capacity (CFM)	Length	Width	Height
4,000	9 ft 0 in	2 ft 6 in	7 ft 0 in
6,000	9 ft 0 in	3 ft 6 in	7 ft 0 in
10,000	10 ft 0 in	4 ft 0 in	7 ft 0 in
14,000	10 ft 0 in	5 ft 6 in	7 ft 0 in
18,000	10 ft 0 in	7 ft 0 in	7 ft 0 in
22,000	12 ft 0 in	6 ft 0 in	7 ft 0 in
26,000	12 ft 0 in	7 ft 0 in	7 ft 0 in
30,000	12 ft 0 in	8 ft 0 in	7 ft 0 in
34,000	14 ft 0 in	7 ft 0 in	8 ft 3 in
40,000	14 ft 0 in	8 ft 0 in	8 ft 3 in
46,000	14 ft 0 in	9 ft 0 in	8 ft 3 in
50,000	16 ft 0 in	9 ft 0 in	8 ft 3 in
60,000	16 ft 0 in	10 ft 0 in	8 ft 3 in
75,000	19 ft 0 in	10 ft 0 in	10 ft 0 in
90,000	22 ft 0 in	11 ft 6 in	11 ft 3 in
105,000	22 ft 0 in	13 ft 6 in	11 ft 3 in
125,000	22 ft 0 in	16 ft 0 in	11 ft 3 in
150,000	22 ft 0 in	19 ft 0 in	11 ft 4 in
175,000	22 ft 0 in	22 ft 3 in	11 ft 4 in
200,000	22 ft 0 in	25 ft 0 in	11 ft 4 in

Source: Riley/Environeering, Schiller Park, IL.

a series of rods placed perpendicular to the gas flow. Water is injected into the gas stream above the rod bank. Water and gas stream pass through the constricted area between the rods and in this process their velocity is substantially increased, promoting gas/water scrubbing.

A series of baffles separate larger particles of moisture from the gas stream, and a final demister, a chevron section, removes the majority of water particles remaining in the gas.

This unit has relatively simple construction. The rods can be readily replaced and they are standard round shapes, requiring no machining, only cutting to size. Dimensions of typical Ventri-rod scrubber systems are shown in Table 13–2.

References and Bibliography

1. Brink, J. Air pollution control with fiber mist eliminators. *Canadian Journal of Air Pollution Control* (June 1963): 134–138.

2. Estimating size and cost of venturi scrubbers. Special Report, *Chemical Engineering* (undated).

3. Mottola, A. Diffusivities streamline wet scrubber design. *Chemical Engineering* (December 19, 1977): 77–80.
4. Adams, A. Corrosion problems with wet scrubbing equipment. *APCA Journal* (April 1976) 26/4: 303–307.
5. Sheppard, W., and G., McDowell. Controlling corrosion in flue gas scrubbers. *Plant Engineering* (February 22, 1979): 127–129.
6. Pierce, R. Estimating acid dewpoints in stack gases. *Chemical Engineering* (April 11, 1977): 125–128.

14

Fabric Filters

Fabric filters are prevalent in all types of industrial applications. Those types of interest in incinerator applications are the baghouse and the high efficiency particulate air (HEPA) filter. Baghouses are essentially a series of permeable bags which allow the passage of gas but not particulate matter. They are effective for particle sizes down to the submicron range. HEPA filters are normally used for particle sizes below 10μ and have excellent removal efficiencies for particle sizes less than 1μ.

Fabric Filter System Considerations

There are a number of objectives in the selection and design of fabric filter systems, or baghouses. First of all, they must provide the desired degree of filtration. This is primarily a function of selecting the proper filtration fabric and establishing an adequate fabric area or filtering velocity to provide the removal efficiency to meet the applicable regulatory requirements.

A second factor is providing the longest possible bag life. There may be hundreds of bags within a single installation and individual bag life can be a major factor in maintenance costs and viability of an installation. The optimum bag life is achieved with selection of optimim gas (filtering) velocity, bag material quality, bag cleaning method, the frequency of cleaning and equipment maintenance.

It is necessary to have the ability to properly clean the bags on an automatic and regular basis. The required cleaning capability can be defined as that minimim cleaning energy necessary to maintain reasonable and stable pressure drops across the bag assembly at all times.

The fourth consideration in good baghouse design is provision of an adequate gas and dust distribution system. Ideally, each square inch of filtering medium should see the same dust loading, particle size distribution, and gas density. The result of this design would be a dust cake of uniform thickness and uniform filtering velocity throughout all areas of the collector. This goal would provide the lowest overall pressure drop and the longest bag life. To achieve this goal the design of the inlet and outlet sections, turning vanes, baffle plates, bag spacing, bag length, and dampers must be carefully controlled.

It is necessary, as a further consideration, to provide effective dust removal from the collector. Once the dust is collected on the filters and is transferred into the dust hopper, it must be removed. Baghouses are not storage vessels and their hoppers should not be used for storage of collected dust. They should be cleaned out on a regular basis, preferably by automated conveyors or other equipment. Full hoppers of dust could cake and harden, can plug, bridge over, and, in extreme cases cause burning of holes in filter bags.

Baghouse Design

Filter fabrics will usually be woven with relatively large spaces, in excess of 50μ. The filtering process, therefore, is not just simple sieving since particles less than 1μ are caught by the bags. Filtering occurs as a result of impaction, diffusion, gravitational attraction, and elecrostatic forces generated by interparticle friction.

As dust collects on a bag filter surface the resistance of the filter to gas flow increases. In addition to this dust mass resistance there is a clean cloth resistance which is a function of the type of cloth fiber and its weave. Clean cloth resistance is measured by an ASTM permeability test procedure: permeability is measured as that air volume, in dry standard cubic feet per minute (DSCFM), that will produce a pressure differential of 0.5 inch WC across 1 square foot of cloth area. The usual values of permeability vary from 7 to 70 DSCFM per square foot.

Typical filter fabrics in current use are listed in Table 14–1. Operating temperatures for continuous (long) and intermittent (short) duty are provided as well as information on flammability, permeability, and resistance to corrosive attack. New filter fabrics are constantly developed for new applications and installations and fabric manufacturers should be consulted for current materials and practices.

The filtering ratio of a fabric is the ratio of gas flow, in cubic feet per minute, to filter area, in square feet. The more efficient the filter cloth, the greater the filtering ratio. Figure 14–1 shows the filtering ratios for various materials as a function of dust loading, grains of particulate per cubic foot of gas. Particulate from an incinerator would most closely resemble carbon black or diatomaceous earth, curves numbered 2 and 5 respectively.

A major feature of a baghouse is its need to discharge collected particulate on a regular basis. A number of different methods have been developed including the following:

- Shaker mechanism, Figure 14–2. An eccentric rod physically shakes a bag section and the falling particles drop to the bottom of a hopper, or silo, by gravity, for eventual removal and disposal. The shaker motor is sequenced to operate in conjuction with operation of the fresh air dampers. Fresh air is admitted into that damper section that has its bags shaken, or agitated. The fresh, clean air aids in discharging the collected dust.

Table 14–1 Characteristics of Fabric Filters

Fiber	Operating Exposure °F		Supports Combustion	Air Permeability	Resistance			
	Long	Short			Abrasion	Mineral Acids	Organic Acids	Alkali
Cotton	180	225	yes	10–20	G	P	G	G
Wool	200	250	no	20–60	G	F	F	P
Nylon	200	250	yes	15–30	E	P	F	G
Orlon	240	275	yes	20–45	G	G	G	F
Dacron	275	325	yes	10–60	E	G	G	G
Polypropylene	200	250	yes	7–30	E	E	E	E
Nomex	425	500	no	25–54	E	F	E	G
Fiberglass	550	600	yes	10–70	P–F	E	E	P
Teflon	450	500	no	15–65	F	E	E	E

Note:

P = Poor, F = Fair, G = Good, E = Excellent
Permeability at 0.5 inch WC

Source: A. Stern, *Air Pollution*, Vol. 3. 2d ed. (New York: Academic Press, 1973).

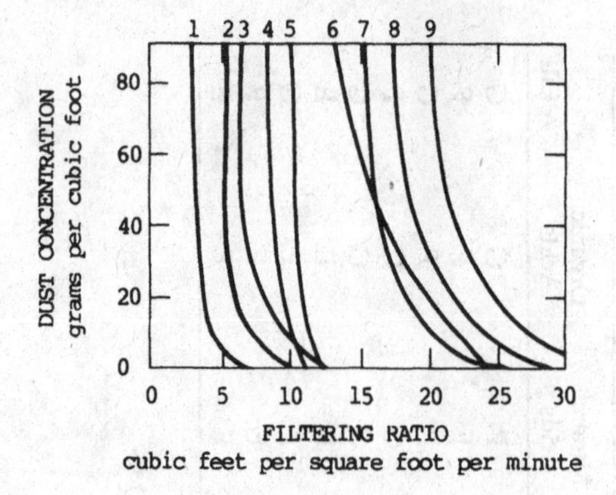

KEY
1. MAGNESIUM TRISILICATE
2. CARBON BLACK
3. STARCH DUST
4. RESINOX
5. DIATOMACEOUS EARTH
6. KAOLIN
7. CEMENT OR LIMESTONE DUST
8. COAL DUST
9. LEATHER BUFFING DUST

FOR NUMBERS 1 THROUGH 6, 99.94-99.99 PERCENT PASSING 325 MESH. FOR NUMBERS 7 AND 8, 95 PERCENT PASSING 200 MESH. NUMBER 9, 60 MESH AVERAGE.

Figure 14–1 Typical filter fabrics.

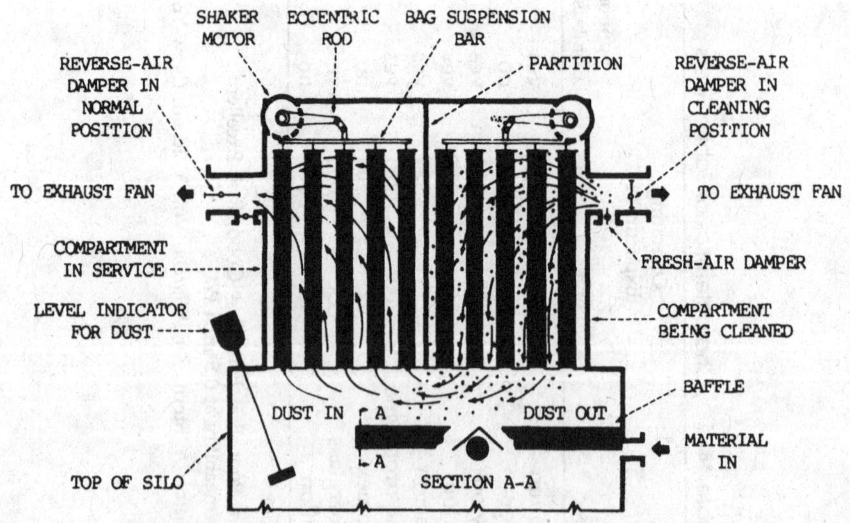

Figure 14–2 Bag filter with shaker mechanism.

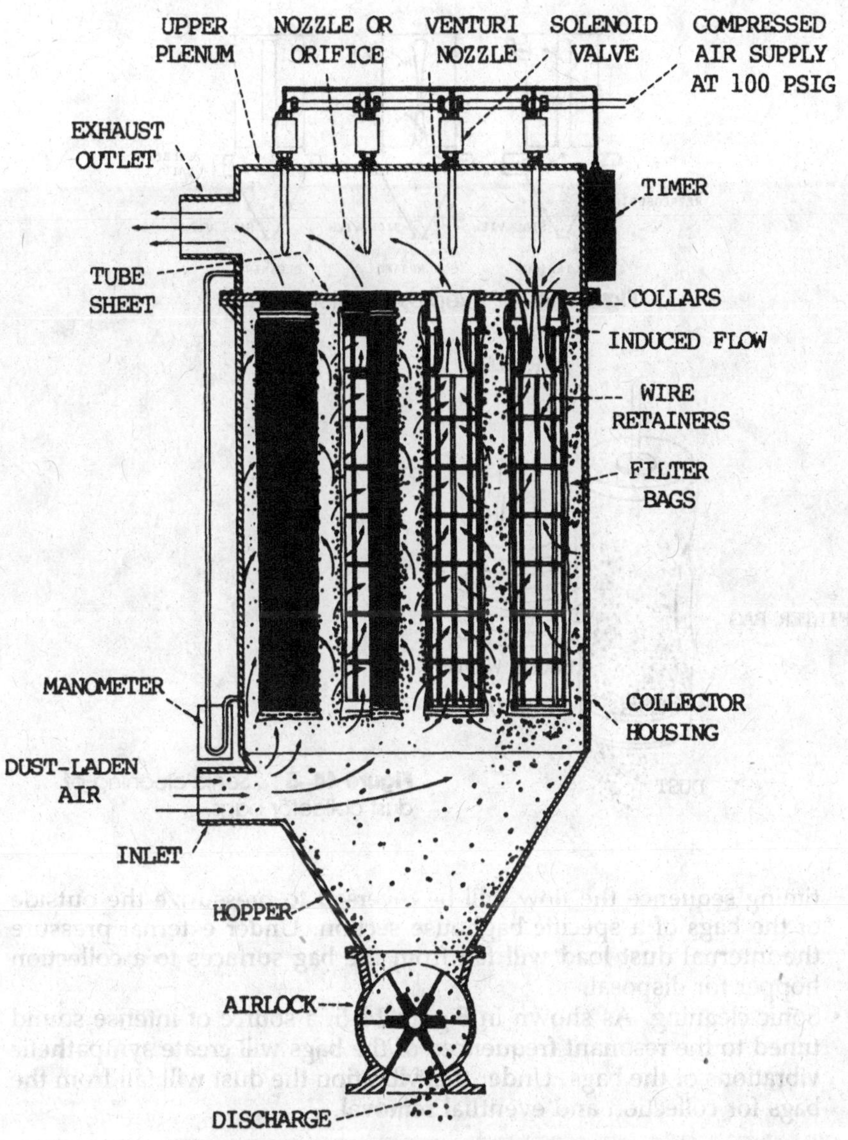

Figure 14–3 Compress air back-wash.

- Compressed air, Figure 14–3. A blast of compressed air is directed into the inside of each bag discharging the dust accumulated on the external surface of the bags. Wire retainers are normally provided to help reinforce the bags against the abrupt action of the air blast.
- Repressurization. The filter sections are independent of each other. Through a series of inlet and exhaust valves, as shown in Figure 14–4, the dirty gas flow passes through the inside of the bags. On a

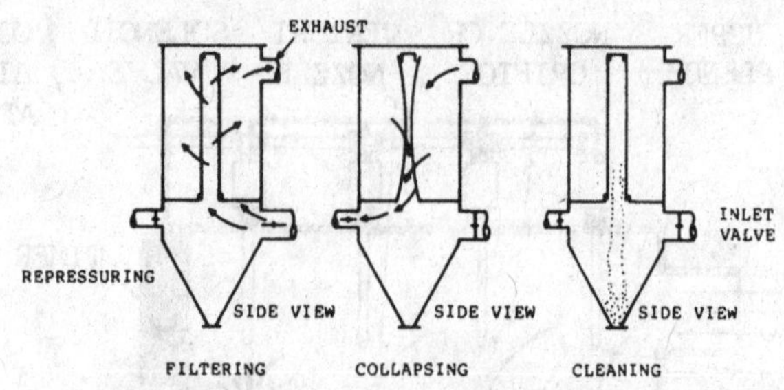

Figure 14–4 Repressurization system.

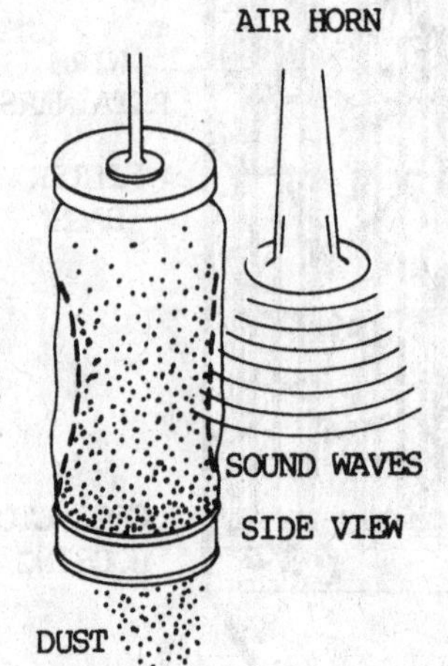

Figure 14–5 Sonic cleaning of dust collector bags.

timing sequence the flow will be reversed to pressurize the outside of the bags of a specific baghouse section. Under external pressure the internal dust load will fall from the bag surfaces to a collection hopper for disposal.

- Sonic cleaning. As shown in Figure 14–5, a source of intense sound tuned to the resonant frequencey of the bags will create sympathetic vibrations of the bags. Under this vibration the dust will fall from the bags for collection and eventual removal.

A typical baghouse assembly is shown in Figure 14–6 with dimensions and operating parameters listed in Table 14–2.

HEPA Filters:

HEPA filters are extremely efficient filters that have been developed for control of particulate from nuclear energy facilities. They can remove over 99.97% of particules of 0.3μ and greater. Their use is not restricted

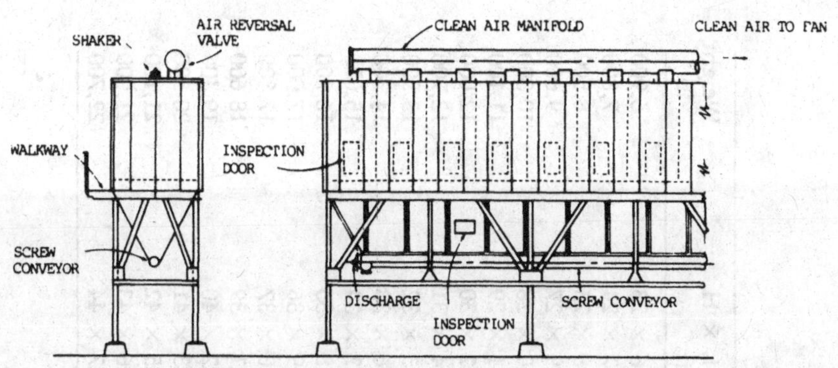

Figure 14–6 Typical baghouse assembly.

to nuclear applications. They are used wherever small particulate removal is required. They are often referred to as "absolute" particulate filters.

Their high removal efficiencies are obtained at the expense of a degree of flexibility. For instance, they will clog when exposed to a wet gas stream. The gas stream must pass through a drying system prior to a HEPA filter if it has a significant moisture content. HEPA filters have, in general, poor resistance to high temperatures and gases must be cooled, often to below 200°F for many applications.

The standard test for evaluating removal efficiency is the dioctylphalate (DOP) smoke test. DOP particles are unique in that they are generated in one size consistently, 0.3μ. The standard test for HEPA filters, or other air pollution control equipment, can therefore reliably be evaluated with the use of this standard size smoke. The efficiency of a filter can be established as the percent removal of DOP smoke with a filter differential pressure drop of 1.0 inches WC.

A typical HEPA filter, as shown in Figure 14–7, is constructed of a glass mat pleated to increase its surface area. The filter is mounted in a frame and a series of frames are mounted in a filter bank to provide the required flow capacity. Table 14–3 lists typical HEPA filter sizes versus flow capacity. Figure 14–8 relates the resistance of a clean HEPA filter to air flow for each of three different size filters.

Dry Scrubbing

Dry scrubbers have been used in the electric power industry, specifically with coal burning power plants, for a number of years. Recently this technology has been adapted to control of incinerator emissions. It is included in this chapter because the dry scrubber system has generally been followed by a baghouse. A particulate collection device is required downstream of the dry scrubber and it is possible, in some instances,

Table 14–2 Typical Baghouse Parameters

Capacity (ACFM) 5:1 Ratio	12:1	Number of bags	Cloth ft2	Number hoppers	W × L × H ft	Weight lb
4,400	10,700	40	888	1	6 × 6 × 23	6,900
5,300	12,800	48	1,066	2	6 × 7 × 24	7,800
6,200	14,900	56	1,243	2	6 × 8 × 25	8,500
7,100	17,100	64	1,421	2	6 × 9 × 27	9,600
8,000	19,200	72	1,598	2	6 × 10 × 28	10,300
8,900	21,300	80	1,776	2	6 × 11 × 29	11,100
9,800	23,400	88	1,954	3	6 × 12 × 30	12,000
10,700	25,600	96	2,131	3	6 × 14 × 31	12,700
11,500	27,700	104	2,309	3	6 × 15 × 32	13,500
12,400	29,800	112	2,486	3	6 × 16 × 33	14,300
13,300	32,000	120	2,664	3	6 × 17 × 34	15,000
14,200	34,100	128	2,842	4	6 × 18 × 35	16,200
15,100	36,200	136	3,019	4	6 × 19 × 36	17,000
16,000	38,400	144	3,197	4	6 × 20 × 37	17,800
16,900	40,500	152	3,374	4	6 × 21 × 39	18,600
17,800	42,600	160	3,552	4	6 × 22 × 40	19,400
18,600	44,800	168	3,730	5	6 × 23 × 41	20,300
19,500	46,900	176	3,907	5	6 × 25 × 42	21,100
20,400	49,000	184	4,085	5	6 × 26 × 43	21,900
21,300	51,100	192	4,262	5	6 × 27 × 44	22,700

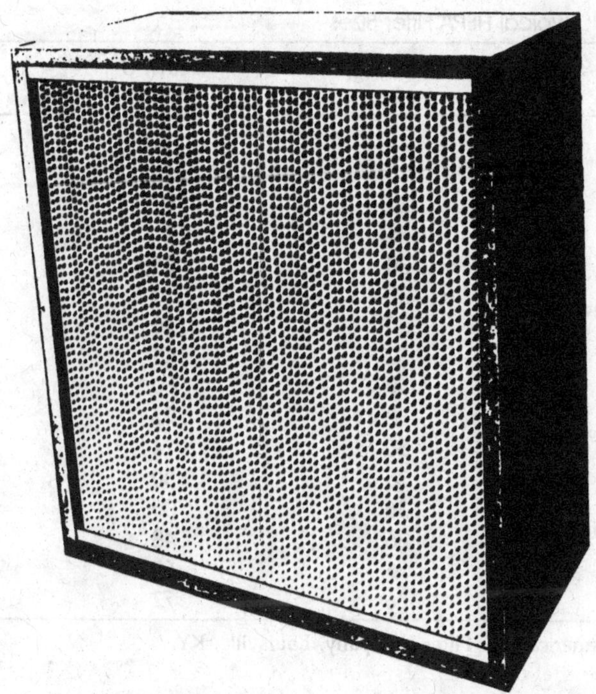

Figure 14–7 HEPA filter. (*Source: American Air Filter Co., Inc., Louisville, KY.*)

to substitute an electrostatic precipitator for the baghouse, but that has, to date, not been done in incinerator gas cleaning system design.

The Teller System, marketed by American Air Filter (Louisville, KY), was the first dry scrubbing system utilized on a refuse incinerator, the Framingham Refuse Reduction Facility. The system consists of two basic components: an Upflow Quench Reactor, Figure 9–9, and a Dry Venturi, Figure 14–10.

Hot incinerator exhaust gas enters the inlet of the quench reactor, a cyclonic element where larger particles drop out of the stream by inertial forces. Rising through the reactor, an alkali solution is sprayed into the swirling gas stream and fully wets it. The alkali (a lime slurry or sodium carbonate solution) neutralizes the acid component of the gas and quenches any sparklers within the gas flow. The reactor is designed to promote neutralization of the acidic gaseous components within 1 second by formation of an alkaline mist within the reactor. The gas residence time within the incinerator is designed for a nominal 7 seconds.

The neutralized gas stream leaves the quench reactor at its adiabatic temperature (normally from 150°F to 180°F) and passes through the dry venturi. A highly crystalline inert material, from 3μ to 20μ particle size, is injected into the gas stream through the venturi. Talc can be used, or waste product fines can be used from various manufacturing industries.

Table 14–3 Typical HEPA Filter Sizes

Rated SCFM at 1 in. W.C.	Height-A Inches	Width-B Inches	Depth-C Inches
900	24	36	5.875
1230	24	48	5.875
1550	24	60	5.875
1900	24	72	5.875
750	30	24	5.875
925	30	30	5.875
1150	30	36	5.875
1550	30	48	5.875
1975	30	60	5.875
2350	30	72	5.875
900	36	24	5.875
1150	36	30	5.875
1400	36	36	5.875
1900	36	48	5.875
2350	36	60	5.875
2850	36	72	5.875

Source: The American Air Filter Company, Louisville, KY.

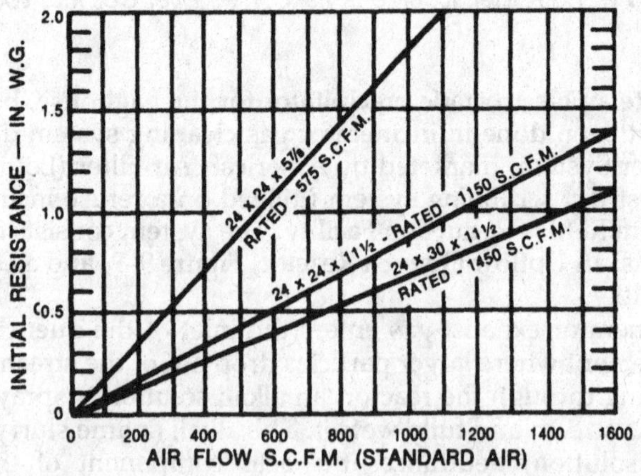

Figure 14–8 Resistance versus air flow for a typical HEPA filter.

Within the venturi the generated turbulence will tend to complete the neutralization process. In addition, the injected powder will adsorb particulate matter within the gas stream and will also act as a catalyst in promoting agglomoration of particulate matter. Particles down to the submicron range have been found to agglomorate to particles of 10μ and greater.

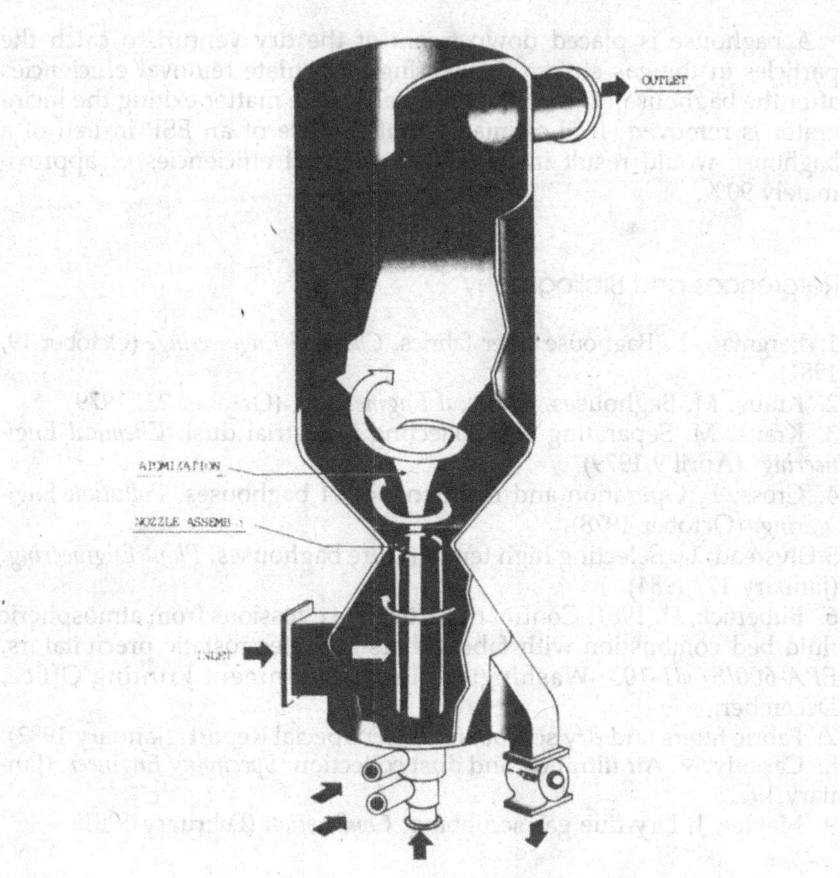

Figure 14–9 Upflow quench reactor.

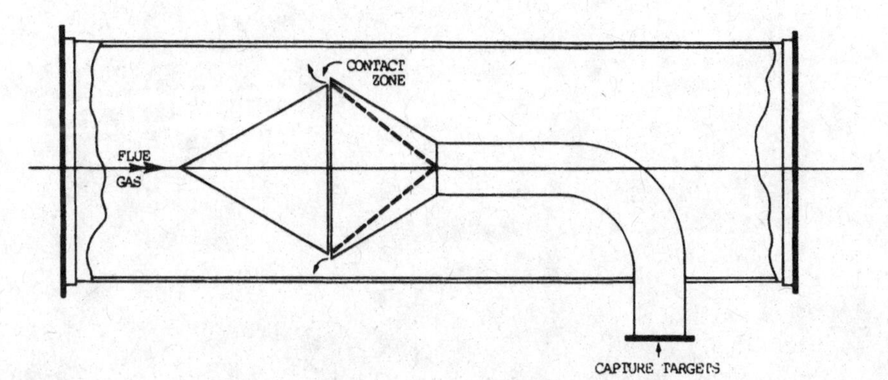

Figure 14–10 Dry scrubber.

A baghouse is placed downstream of the dry venturi to catch the particles in the gas stream. Measuring particulate removal efficiencies after the baghouse, 95 to 99% of the particulate matter exiting the incinerator is removed. It is estimated that the use of an ESP in lieu of a baghouse would result in particulate removal efficiencies of approximately 90%.

References and Bibliography

1. Bergman, L. Baghouse filter fabrics. *Chemical Engineering*. (October 19, 1981).
2. Kraus, M. Baghouses. *Chemical Engineering*. (October 23, 1979).
3. Kraus, M. Separating and collecting industrial dust. *Chemical Engineering*. (April 9,1979).
4. Cross, F. Operation and maintenance of baghouses. *Pollution Engineering*. (October 1978).
5. Ulvstead, U. Selecting high temperature baghouses. *Plant Engineering*. (January 12, 1984).
6. Bubenick, D. 1981. Control of particulate emissions from atmospheric fluid bed combustion with fabric filters and electrostatic precipitators, *EPA-600/S7-81-105*. Washington, DC; Government Printing Office, December.
7. Fabric filters and dry scrubbers. *Power* Special Report. (January 1982).
8. Cassidy, V. Air filtration and dust collection. *Specifying Engineer*. (January 1982).
9. Meyler, J. Dry flue gas scrubbing. *Combustion* (February 1981).

15

Electrostatic Precipitators

Electrostatic precipitators (ESPs) are effective devices for the removal of airborne particulate matter. The first commercial ESP was designed and constructed by Frederick G. Cottrell earlier in this century and they are sometimes referred to as Cottrell process equipment. In Europe they are called electrofilters.

There are three variations of the ESP used for incinerator emissions control: the conventional ESP, the wet ESP, and the Electroscrubber©.

Conventional ESP

An ESP (a typical unit is shown in Figure 15–1) operates as follows:

- The gas stream passes through a series of discharge electrodes. These electrodes are negatively charged in the range of 1000 to 6000 volts. This voltage creates a corona around the individual electrodes. A negative charge is induced within the particulate matter passing through the corona.
- A grounded surface, or collector electrode, surrounds the discharge electrodes. Charged particles will collect on the grounded surface. These surfaces are usually flat plates.
- Particulate matter is removed from the collector surface for ultimate disposal.

Typical discharge electrode and collector plate designs are shown in Figure 15–2. A variation of this design is the two-stage ESP, where the gas passes through a corona discharge prior to entering the collector plate area, as shown in Figure 15–3.

The ESP is extremely efficient in the collection of small size particulate, down to the submicron range. It can be designed for temperatures as high as 700°F however its efficiency is sensitive to variations in flue gas temperature and humidity. As shown in the graph in Figure 15–4, ESP efficiency will vary substantially when operated either above or below its design point.

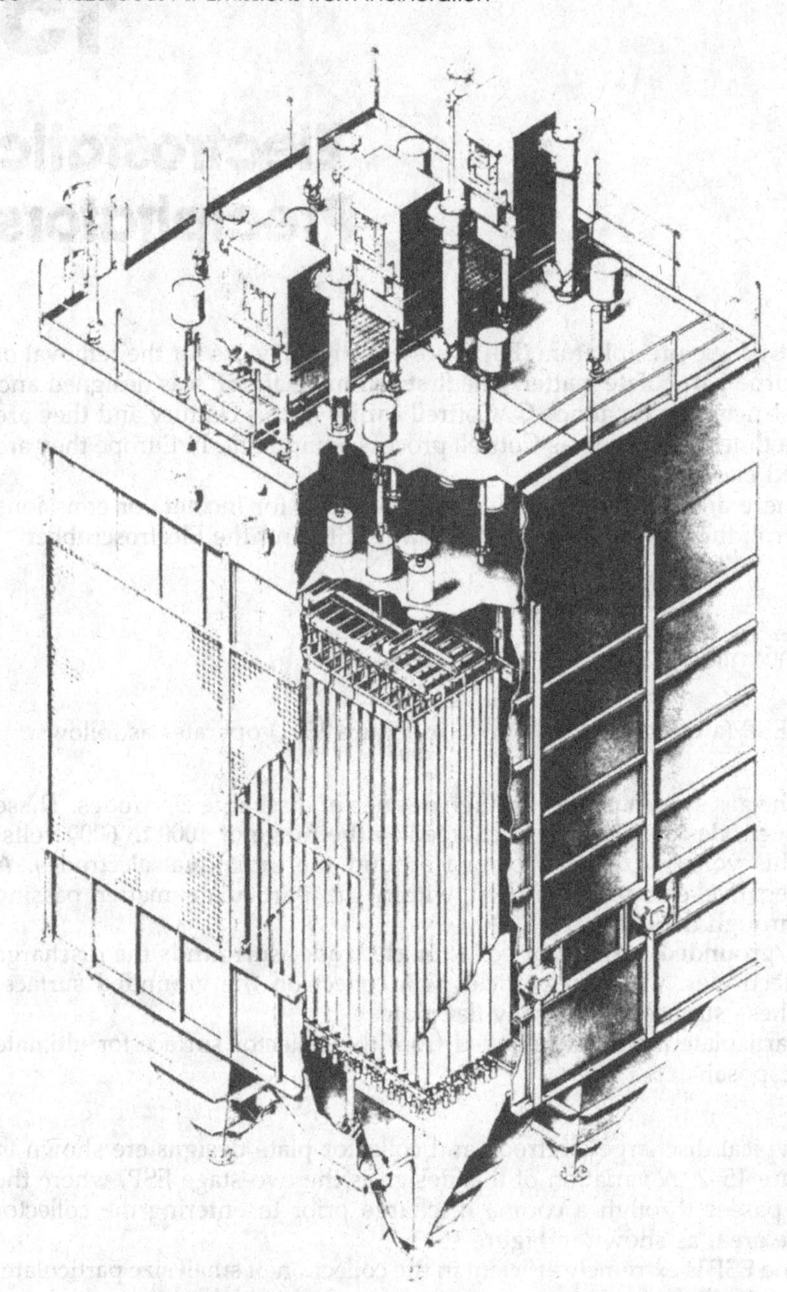

Figure 15–1 Typical electrostatic precipitator.

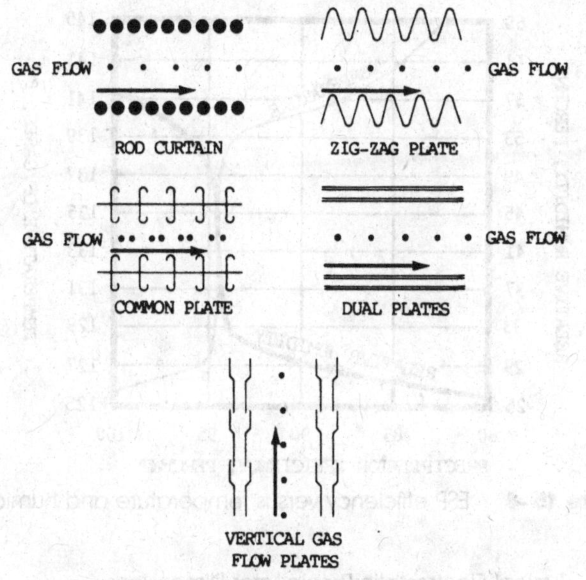

Figure 15–2 Electrodes and Collector Plates.

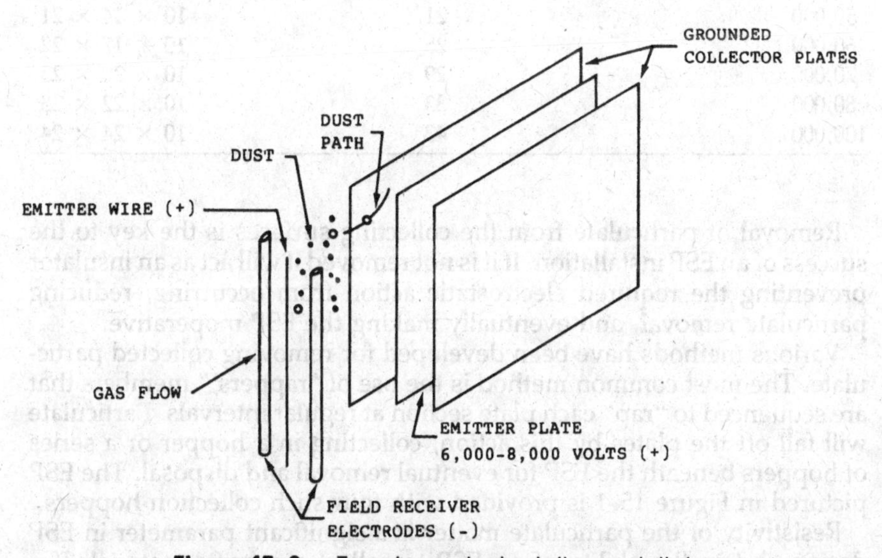

Figure 15–3 Two-stage electrostatic precipitator

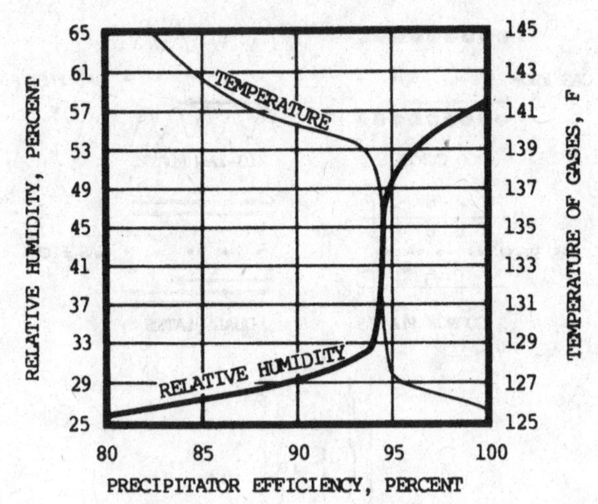

Figure 15–4 ESP efficiency versus temperature and humidity

Table 15–1 Typical Electrostatic Precipitator Dimensions

Capacity—ACFM	No. of Modules	W × L × H (ft)
12,000	5	10 × 4 × 18
20,000	8	10 × 6 × 19
25,000	11	10 × 8 × 19
35,000	14	10 × 10 × 20
40,000	17	10 × 12 × 20
50,000	21	10 × 14 × 21
60,000	25	10 × 17 × 22
70,000	29	10 × 20 × 23
80,000	33	10 × 22 × 23
100,000	43	10 × 24 × 24

Removal of particulate from the collecting surfaces is the key to the success of an ESP installation. If it is not removed it will act as an insulator preventing the required electrostatic action from occurring, reducing particulate removal, and eventually making the ESP inoperative.

Various methods have been developed for removing collected particulate. The most common method is the use of "rappers," members that are sequenced to "rap" each plate section at regular intervals. Particulate will fall off the plates by this action, collecting in a hopper or a series of hoppers beneath the ESP for eventual removal and disposal. The ESP pictured in Figure 15–1 is provided with four such collection hoppers.

Resistivity of the particulate matter is a significant parameter in ESP design and in the ability of an ESP to collect a specific material. If a particle, or dust, has a high resistivity it is unable to give up its electric charge to the collecting electrode. The dust will therefore build up on the collector, acting as an insulating layer. As this layer increases in depth its surface will develop a significant electric charge relative to the collecting electrode. Eventually the insulating dust layer will prevent the ESP from collecting particulate.

With too low a resistivity a dust will readily relinquish its negative charge to the collector and will assume a positive charge. With the collecting electrode at a positive potential, the particle is repelled back to the gas stream. In the gas stream, which is saturated with negatively charged particles, the dust will pick up a negative charge again and will eventually return to the collector plate and be repelled. The low resistivity particle, therefore, will be successively repelled by the collecting electrode and will not be collected. It will pass through the ESP system.

Electrical resistivity is measured in units of ohm-centimeters. The optimum resistivity for particulate matter to be effectively collected within an ESP is from 10,000 to 10 billion ohm-centimeters. The resistivity of most materials varies significantly with temperature and the use of an ESP may well dictate the temperature range of collection, that temperature range that the gas must be maintained at so that its resistivity is within acceptable limits.

With the proper resistivity a dust particle will relinquish part of its charge to the collecting electrode. The rate at which the charge dissipates increases as the dust layer builds up. When the weight of the collected dust exceeds the electrostatic force available to hold the layer to the collector, the dust particles will fall off under their own weight or will be jarred loose when the collectors are rapped.

Another particulate quality asociated with resistivity is the tendency to agglomorate, to form a hard or tarlike mass impossible to rap free. The tendency to agglomorate is influenced by the quantity of moisture within the gas stream and the temperature of the stream.

Collection is a function of gas velocity as well as the other factors noted. The velocity through the collector plates will normally range from 2 to 4 feet per second.

Dimensions of a typical conventional electrostatic precipitator system are listed in Table 15-1, as a function of inlet gas flow.

Wet Precipitators

A variation of the ESP is the wet precipitator. This unit normally has a preconditioning section where the entering gas is sprayed with water to reduce its temperature, remove larger particles, and to provide some acid gas absorption. The collection surface is kept wetted where liquid is used to continuously wash away impacted particulate. A typical wet ESP is shown in Figure 15–5. It is a tubular unit with dirty gas entering its top, circulating throughout the ESP and directed up to a top discharge.

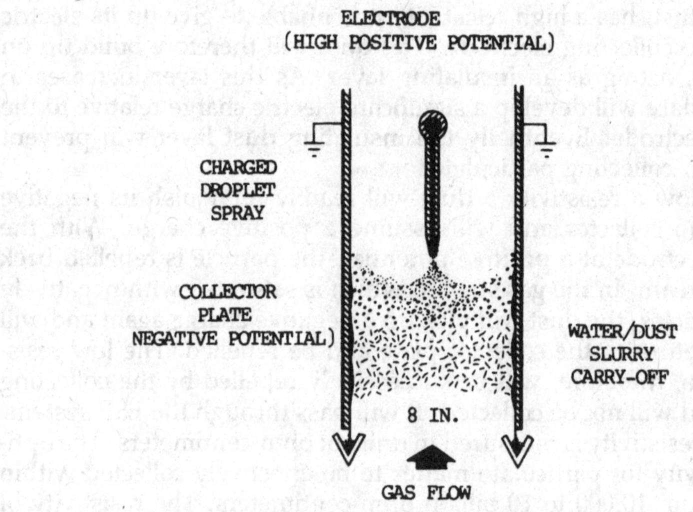

Figure 15–5 Wet Electrostatic precipitator.

Particle collection occurs by the introduction of evenly distributed liquid droplets in the gas stream through sprays located above the electrostatic field sections and subsequent migration of the charged particles and liquid droplets to the collecting plates. Figure 15–6, a cross section through a wet ESP, illustrates this action.

The liquid droplets tend to collect on the plates to form a downward flowing film keeping the plates clean. The last section of the wet ESP is often operated dry, designed with collection baffles, to control the carryover of liquid droplets and mists to downstream equipment.

As with all control equipment operating in a wet mode, the wet ESP is subject to acid corrosion. The water flow will become acidic and it has the potential to create corrosive problems within the ESP and in connected ductwork, piping, and tanks.

Electroscrubber©

The Electroscrubber© (a product of the Combustion Power Company, Menlo Park, CA) has been found to be effective in emissions control from municipal solid waste incinerators. Figure 15–6 is a cut through the unit. Hot, dirty gas passes through the outer of two concentric louvered tubes. The annular space between the tubes is filled with pea-sized gravel media. The gas is distributed through the media and exits the internal louvered cylinder cleaned of the majority of its particulate load.

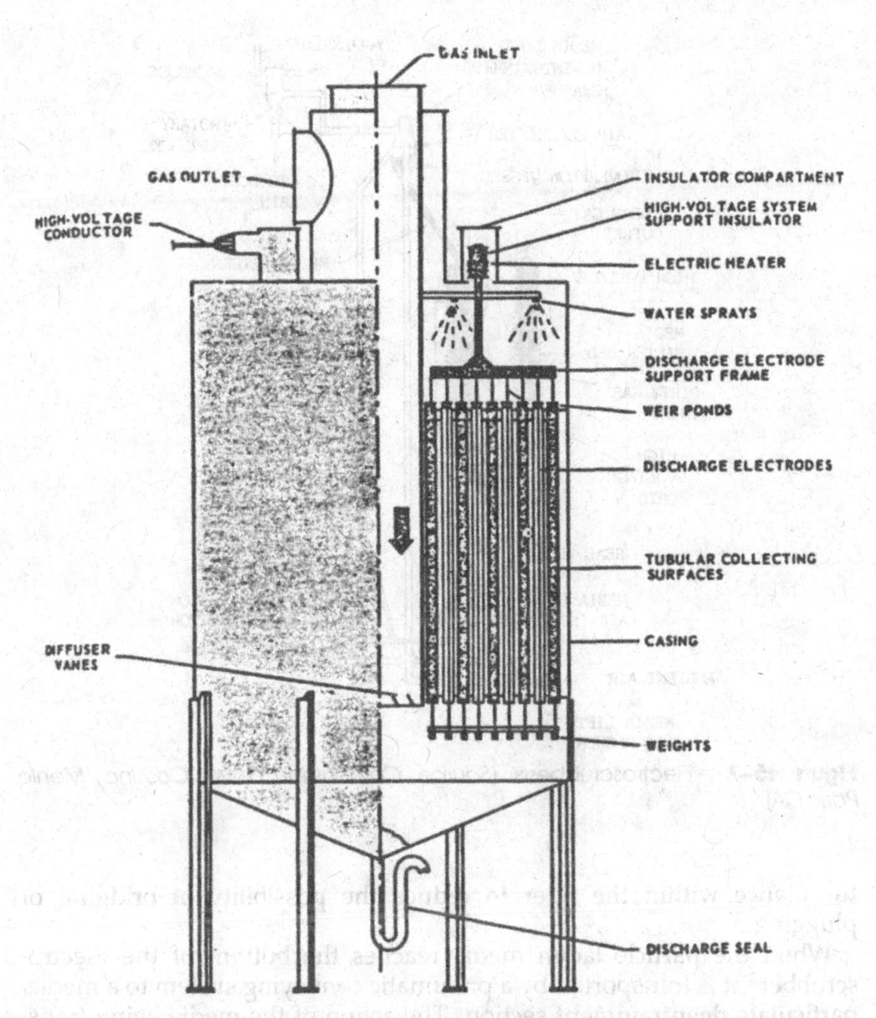

Figure 15–6 Tubular wet Electrostatic precipitator. (*Source: Joy Manufacturing Co.*)

An electric field of up to 50,000 volts is applied to the gravel media with a cagelike structure within the louvered cylinders. All particles produced in industrial processes have a slight positive or negative charge. This electrical field is sufficiently high to force the particulate matter within the gas stream (with its electrical charge) to migrate either to or from the individual filter granules. The gas particles are captured by the granular material, removed from the flowing gas stream.

The gas velocity entering the outer louver ranges from 100 to 150 feet per minute. The media is continually moving downward from the louver at a rate of 6 to 10 feet per hour and in doing so generates enough

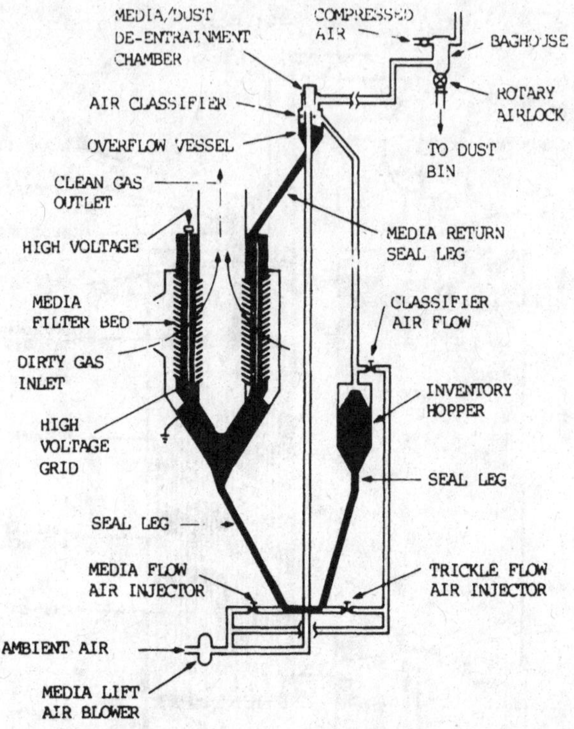

Figure 15–7 Electroscrubber® (*Source: Combustion Power Co., Inc., Menlo Park, CA*)

turbulence within the filter to reduce the possibility of bridging or plugging.

When the particle laden media reaches the bottom of the Electroscrubber© it is transported by a pneumatic conveying system to a media/particulate deentrainment section. The action of the media being transported vertically in the pneumatic lift pipe separates the particulate from the media so that the particulate can be pneumatically removed from the deentrainment section. The cleaned media drops by gravity from the deentrainment chamber to the louvers for reuse and recycling. The particulate is collected in a dust bin and/or a baghouse for removal from the system and eventual disposal.

This system has been found to be effective for varying exhaust gas flow rates and has high catch efficiencies for submicron particles. There are relatively few moving parts and except for addition or relacement of gravel there is little maintenance required.

There are few of these units in use in incineration facilities. With this limited operating experience there is concern that the flowing gravel may tend to erode the gravel containment section.

References and Bibliography

1. Bubenick, D. 1981. Control of particulate emissions from atmospheric fluid bed combustion with fabric filters and electrostatic precipitators. *EPA-600/S7-81-105*. Washington, DC: Government Printing Office, December.
2. Electrostatic precipitation. *Chemical Engineering Progress* Special Report (April 1966).
3. Schneider, G. Selecting and specifying electrostatic precipitators. *Chemical Engineering* (May 26,1975).
4. Jaros, W. FRP for ESP's. *Pollution Engineering*. (June 1982).

16

Control System Selection

There are several important factors governing the selection of a system and equipment for controlling incinerator emissions. Regulatory requirements, the nature of the gas stream and utility availability must all be considered.

Regulatory Requirements

The first step in control system selection is a determination of the relevant statutory requirements. As noted in previous chapters, governmental regulations can be extremely complex. Often there is uncertainty in the applicability of these regulations.

Where a new incinerator is considered for an installation, the nature of the installation is a major factor in defining incinerator emissions parameters. If the existing facility is not a "major source," as defined by the National Ambient Air Quality Standards (NAAQS), the addition of an incinerator on site may add significant emissions to the total facility discharge to place it in the "major source" category. An incinerator may be large enough to constitute a major source of emissions by itself. As noted in the chapter on the NAAQS, the procedure mandated by this statute is sufficiently rigorous and complex to veto incinerator construction if new source review procedures will be necessary.

By proper choice of emissions parameters, the incinerator discharge may be able to be brought below the level where the incinerator, or the existing facility, is considered a major source of air pollutants.

Other federally mandated requirements, such as those for refuse or sewage sludge incinerators, or for hazardous waste incinerators, are also factors which may prove critical for control system selection. State statutes, as defined in chapter 2 "Statutory Requirements," also have to be considered when determining the allowable emissions levels from an incinerator discharge.

The first consideration in establishing incinerator discharge parameters is, therefore, the impact of governmental statutes on the proposed incinerator location. A corollary is that, except for relatively small incinerators, an incinerator exhaust system selection or design should not proceed until the location of the incinerator is established. Once the site, or choice of sites, is known, the following factors can be evaluated:

- Is the location an attainment or nonattainment area for any of the criteria pollutants?
- What is the existing level of pollutant discharge, if any, at the proposed site?
- Has a new source review been performed for existing equipment at the proposed site?
- What are the state regulations?

The most helpful people in determining the above parameters are normally local regulatory agency personnel. They have intimate knowledge of the statutes and statutory classifications within their cognizant areas. Consulting with them early in a project and seeking their advice in helping to set emissions parameters can only help in obtaining the necessary reviews and approvals in an expeditious manner.

Utility Availability

A wet control device uses a relatively large quantity of water for effective operation. If water is not available, a wet device cannot be used, regardless of other considerations. Where water is available, use of a wet scrubbing device is often a preferred option.

Wet scrubbing systems require, in addition to a supply of water, water at a low enough temperature for effectively cooling the gas stream. It is often desireable to reduce the temperature of the gas exiting the stack and the use of relatively cool water may be necessary.

An electrostatic precipitator will not need a supply of water for operation, but it may produce a discharge that must be cleaned before being sewered. Normally a source of electrical power is brought to the electrostatic precipitator (ESP) and the ESP will be provided with its own equipment to convert input AC to regulated DC.

The sewerage requirement may be a limitation on system selection. If a discharge cannot be fed directly to an existing sewer the cost of treating it may be prohibitive.

Predicting Performance

The collection efficiency of a control device is specific to the particular equipment in question, and usually varies from one manufacturer to another. Listed data on system efficiencies can be used to obtain a reasonable estimate of system performance and can be useful in system selection.

The data in Table 16–1 compare the average removal efficiencies for individual gaseous contaminants of various types of control equipment. The relatively high removal of HC1 is a function of the water circulated within the indicated device. The basic design of this equipment does not directly contribute to its ability to remove gaseous components from the incinerator discharge.

Table 16–2 lists average particulate removal efficiency of control equipment as a function of particle size. Below 1μ, which is not included within this data, removal efficiency of all but the electrostatic precipitator (ESP) will show a substantial decrease.

It is usually not possible to obtain data on particle size distribution without performing extensive tests on the equipment in question, burning the fuel (waste) in question. If this information were available control system selection would be a simple exercise, mathcing particle size input to the known incinerator particulate removal efficiency. Normally, the actual selection of a control system, and system operating parameters, require an educated guess on the part of the engineer and the manufacturer precisely because the particle size distribution is not known. Tables 16–3 and 16–4 list particle size distribution for tests on two incinerators, burning sewage sludge and firing refuse, respectively.

Selection

As noted above, there are many factors to consider in selection of the appropriate control device. The nature of the waste stream is a factor, as well as the type of incinerator equipment that is used. In addition, regulatory constraints must be considered

In the last analysis, the selection of the correct system depends upon experience with similar applications, aided by past performance as represented by the tabular information included within this chapter.

References and Bibliography

1. Ondov, J. Elemental emissions from a coal fired power plant: Comparison of a venturi wet scrubber system with a cold side electrostatic precipitator. *Environmental Science and Technology* (May 1979).
2. USEPA. 1971. The Federal R and D plan for air pollution control by combustion process modification. *EPA/CPA 22-69-147*. Washington, DC: Government Printing Office, January.
3. USEPA. 1978. Industrial guide for air pollution control. *EPA 625/6-78-004*. Washington, DC: Government Printing Office, June.
4. Greiner, G., and J. Szalay. Latest developments in air pollution control technology. *Specifying Engineer* (September 1981).

Table 16–1 Average Control Efficiencies of Air Pollution Control Systems

System Type	Particulate Mineral	Particulate Combustible	Hydrogen Chloride	Nitrogen Oxides	Volatile Metals	Sulfur Oxides
None (settling chamber)	20	2	0	0	2	0
Dry Expansion Chamber	20	2	0	0	0	0
Spray Chamber	40	5	40	25	5	0.1
Wetted Wall Chamber	35	7	40	25	7	0.1
Wetted Close-Space Baffles	50	10	50	30	10	0.5
Mechanical Cyclone (dry)	70	30	0	0	0	0
Medium Energy Wet Scrubber	90	80	95	65	80	1.5
Electrostatic Precipitator	99	90	0	0	90	0
Fabric Filter	99.9	99	0	0	99	0

Notes:
1. Removal efficiencies for carbon monoxide and hydrocarbons are zero for the above systems
2. Combustible particulate and volatile metals are assumed to be primarily $< 5\mu$ mean diameter

Source: K. Akita, *Journal of Polymer Science*, 5: (1978), 833.

Table 16–2 Average Collection Efficiencies of Gas Cleaning Equipment

Equipment Type	Percentage Efficiency 50μ	5μ	1μ
Inertial Collector	95	16	3
Medium Efficiency Cyclone	94	27	8
High Efficiency Cyclone	96	73	27
Impingement Scrubber	98	83	38
Electrostatic Precipitator	>99	99	86
Wet Electrostatic Precipitator	>99	98	92
Flooded Disc Scrubber, Low Energy	100	99	96
Flooded Disc Scrubber, Medium Energy	100	99	97
Venturi Scrubber, Medium Energy	100	>99	97
Venturi Scrubber, High Energy	100	>99	98
Shaker Type Fabric Filter	>99	>99	99
Reverse Jet Fabric Filter	100	>99	99

Source: J. Stairmand, *Chemical Engineering* (London), 194: (1979), 310.

Table 16–3 Sewage Sludge Incineration Airborne Particle Size

		Percent by Weight Less Than Indicated Size	
		Location 1[a]	Location 2[b]
Particulate Loading,	grains/dscf	1.88	0.01
	pounds/hour	217.01	1.48
	pounds/wet ton	52.08	0.36
Size Distribution, microns			
	18.7	37.9	100.0
	11.7	30.6	98.0
	8.0	16.4	94.9
	5.4	6.6	93.4
	3.5	2.6	92.8
	1.8	1.6	83.3
	1.1	0.9	67.7
	0.76	0.1	54.6

[a]Measured at incinerator outlet, without controls, burning 100 tons/day wet sludge cake.
[b]Measured after venturi scrubber with a total pressure drop of 30 inches WC.

Source: C. R. Brunner, *Incineration Systems: Selection and design*. (New York: Van Nostrand Reinhold, 1984).

Table 16–4 Refuse Incineration Airborne Particle Size

	Unit	
	250 Ton/day	120 Ton/day
Particle Specific Gravity, lb/cf	2.70	3.77
Particle Bulk Density, lb/cf	30.9	9.4
Loss On Ignition ∝ 1400°F, %	8.2	30.4
Size Distribution, % By Weight Less than indicated size:		
microns		
30	40.4	50.0
20	34.6	45.0
15	31.1	42.1
10	26.8	38.1
8	24.8	36.3
6	22.3	33.7
4	19.2	30.0
2	14.6	23.5
Particulate Emission Rate:		
lb/ton	24.6	9.1
lb/hour	256.3	45.5

Note: Measurements made at incinerator exit, prior to any control equipment.
Source: C. R. Brunner, *Incineration Systems: Selection and design*. (New York: Van Nostrand Reinhold, 1984), p. 316.

5. Calvert, S. How to choose a particulate scrubber. *Chemical Engineering* (October 29,1977).
6. Weinstein, N., and R. Toro. Control systems on municipal incinerators. *Environmental Science and Technology* (June 1976).
7. Stevens, J. Energy recovery and emissions from municipal waste incineration. *Ontario APCA Journal* (September 1981).

17

Dispersion Calculations

The discharge from a stack will travel through the air, from the source to a receptor. The characteristics of this discharge is a major consideration in the evaluation of the source, the incineration facility.

Consider that a discharge will contain gas components, and aerosols, both solid and liquid. It will normally be forced out of the stack by a fan, or in the absence of a fan it will have a lower velocity from natural draft.

The source will normally have relatively small dimensions, and as the discharge travels through the air, away from the confines of a stack or a chimney, it will disperse and occupy greater and greater volume.

Not only are the original characteristics of the discharge important in a determination of the dispersion profile, but climatology and topography must be considered. The disperson characteristics of a discharge are dependent upon the ambient air temperature and pressure, presence of sunlight, clouds, or absence of sunlight (night), and the wind speed and direction. With regard to topological parameters, the discharge from a source on a mountain will not disperse in the same manner as if it were in a valley or on a flat plain. The presence of buildings or other structures in the vicinity of the source, or the receptor, is also a factor in the determination of actual dispersion.

An accurate calculation of dispersion includes detailed information on discharge, climatology, and receptor conditions. The amount of information required is normally well beyond the ability of an unaided human being to accumulate and analyze. Computer models have been developed to generate a dispersion profile from the collected data; however, running these models can be relatively expensive and requires highly specialized personnel and equipment.

Many techniques and handbooks have been developed and published to approximate the dispersion characteristics of a discharge. One of the first, widely used handbooks, written by D. Turner,[1] provides simplified but conservative forecasts of pollutant concentration at a receptor from a source (such as the stack of an incineration system). The techniques for calculating atmospheric dispersion that are presented in this chapter are based on Turner's methods. His methods have been incorporated into the Environmental Protection Agency's (EPA's) UNAMAP model series, a collection of computerized dispersion models (31 different models as of August 1983—UNAMAP Version V) which are capable of simulating dispersion for a variety of source types and environmental settings.

It must be remembered that the following calculations represent an approximation of a fairly detailed and lengthy procedure. Surface effects such as the presence of buildings or trees between the source and the receptor, atmospheric conditions such as precipitation or excessive humidity, or source conditions such as intermittent operation are not considered. In general, however, the simplified approach in this analysis provides higher receptor concentrations than exist in fact. Where these values approach critical values a more detailed analysis will have to be performed. In essence, these calculations will provide a "first-cut" approach to the severity of source conditions on a receptor.

Figure 17–1 defines the frame of reference for the following calculations. The Z-axis is the vertical axis, coincident with the discharge. The X-axis is the parallel to the prevailing wind direction, positive-downwind from the discharge. The Y-axis is a horizontal axis perpendicular to the prevailing wind direction. The "bell" curves superimposed on the dispersing discharge represent the method of calculation for determining the dispersion, or spread, of the discharge as it travels away from the stack. These curves are "normal," or "Gaussian," distribution curves, exponential functions representing natural processes.

System of Units

The units commonly used in dispersion calculations are metric. A set of conversion tables from metric to English, and English to metric is included within the appendix.

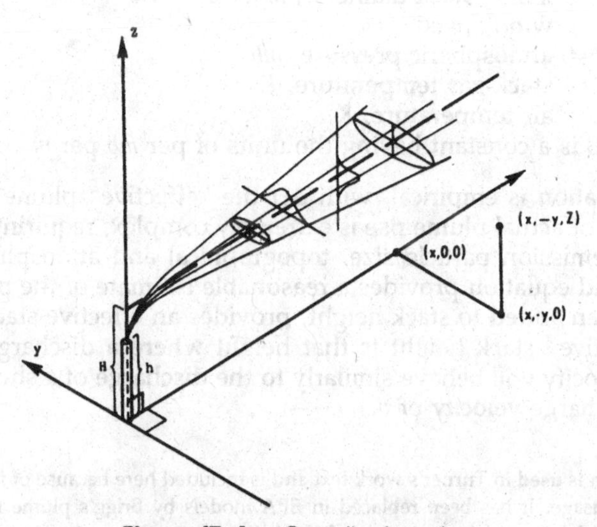

Figure 17–1 Coordinate system

Units used in this chapter and their metric abbreviations are as follows:

Measurement	English Unit	Metric Unit
Weight	pound (lb)	gram (g)
Length	inch (in.)	meter (m)
Length	foot (ft)	meter (m)
Distance	Mile (mi)	kilometer (km)
Pressure	inch WC (in. WC)	millibar (mb)
Pressure	inch Hg (in. Hg)	millibar (mb)
Pressure	psi	millibar (mb)
Temperature	Fahrenheit (f)	Kelvin (K)
Time	second (sec)	second (s)

Effective Height of Source

A discharge will usually exit a stack with a particular velocity. At some height the discharge, or plume, will become essentially level, losing its vertical velocity. The height at which this occurs is the effective stack height, *H*. An equation derived from an equation developed by J. Holland[2] describes the effective stack height, as follows:*

$$\Delta H = \frac{v \cdot d}{u} \left(1.5 + 0.00268 \cdot P \cdot \frac{T_s - T_a}{T_s} \cdot d\right)$$

where ΔH = the rise of the plume above the stack, *m*
v = stack gas exit velocity, *m/s*
d = inside stack diameter, *m*
u = wind speed, *m/s*
P = atmospheric pressure, *mb*
T_s = stack gas temperature, *K*
T_a = air temperature, *K*
and 0.00268 is a constant having the units of per *mb* per *m*.

This equation is empirical, with ΔH the "effective" plume rise. The calculation of actual plume rise is extremely complex, requiring detailed inputs of emission particle size, topographical and atmospheric data. The Holland equation provides a reasonable estimate of the plume rise which, when added to stack height, provides an effective stack height. The "effective" stack height is that height where a discharge of zero vertical velocity will behave similarly to the discharge of a shorter stack with a discharge velocity of v.

*This equation is used in Turner's work text and is included here because of its simplicity and ease of usage. It has been replaced in EPA models by Brigg's plume rise formula which, although more accurate, is more difficult to use.

Table 17–1 Stability Categories

Surface Wind Speed*		Day			Night	
		Incoming Solar Radiation			Thinly Overcast	
m/sec	mph	Strong	Moderate	Slight	>4/8 Low Cloud	>3/8 Cloud
<2	<4	A	A–B	B	—	—
2–3	4–7	A–B	B	C	E	F
3–5	7–11	B	B–C	C	D	E
5–6	11–14	C	C–D	D	D	D
>6	>14	C	D	D	D	D

Note: The neutral class, D, should be assumed for overcast conditions during day or night.

*Wind speed measured at a height of 10 meters above ground.

Source: D. Turner, *Workbook of Atmospheric Dispersion Estimates* (Washington, DC: U.S. Department of Health, Education, and Welfare, Public Health Service, Washington, DC: Government Printing Office, 1969).

The Diffusion Equation

The concentration of an aerosol or particulate has been found to have a gaussian distribution downstream from its source. The equation developed by Turner[1] to determine a concentration downstream from a source includes, therefore, the standard deviation in both the horizontal and vertical planes related to distance and stability category. The stability categories used by Turner are listed in Table 17–1 and are a function of atmospheric conditions; for example, surface wind speed, day or nighttime, and cloud cover.

Turner's diffusion equation is as follows:

$$\omega\,(x,y,z,H) = \frac{Q}{2\pi\sigma_y\sigma_z u} \cdot \exp\left\{-\frac{1}{2}\,(y/\sigma_y)^2\right\} \cdot (A + B)$$

with:

$$A = \exp\left\{-\frac{1}{2}\,\frac{(z - H)^2}{\sigma_z}\right\}$$

$$B = \exp\left\{-\frac{1}{2}\,\frac{(z + H)^2}{\sigma_z}\right\}$$

where: ω = aerosol concentration at stack, g/m^3

Q = rate of discharge, g/s

y = distance from source centerline along the Y-axis as shown in Figure 17–1

z = height above ground, m
H = effective stack h eight, $H = h + \Delta H$, m
h = actual stack height, m
ΔH = effective plume rise, from Holland's equation, m
σ_y = standard deviation perpendicular to the prevailing wind direction, along the horizontal Y-axis, per the values in Figure 17–2, m
σ_z = standard deviation in the vertical direction, along the Z axis, per the values in Figure 17–3, m
$\sigma_y\sigma_z$ = the product of σ_x and σ_y, as per Figure 17–4, m^2.

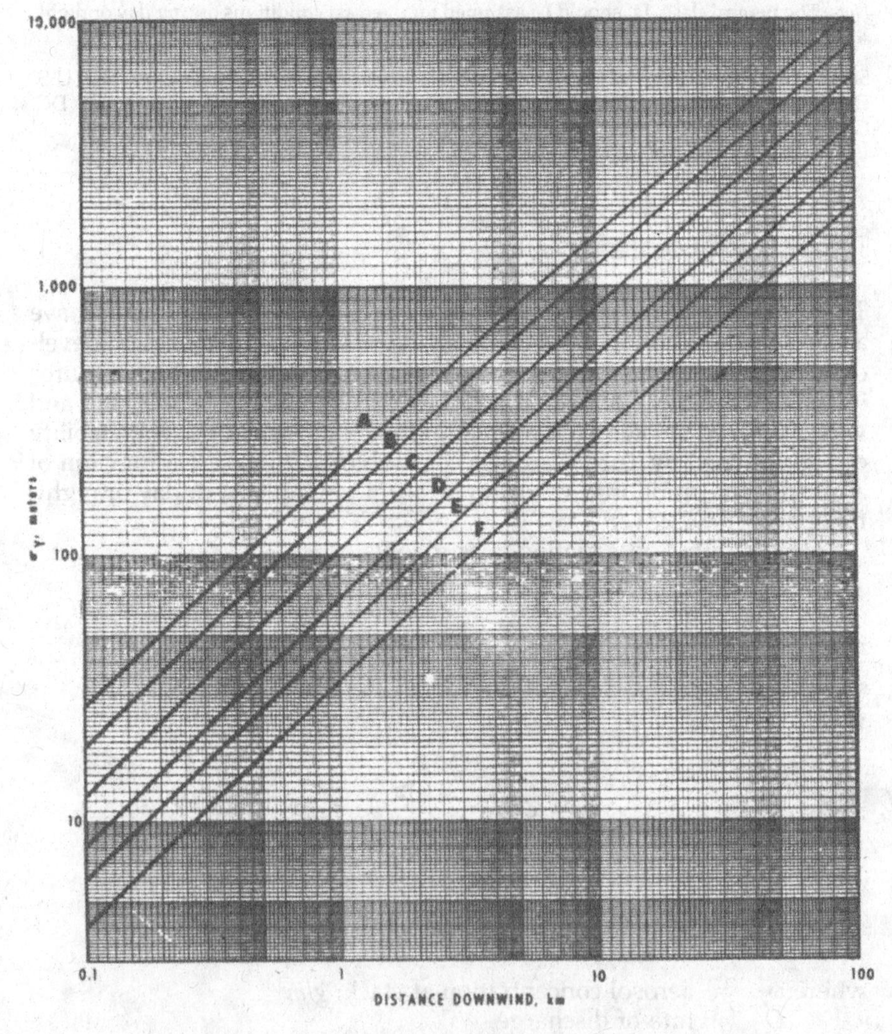

Figure 17–2 Horizontal dispersion coefficient as a function of downwind distance from the source and stability category.

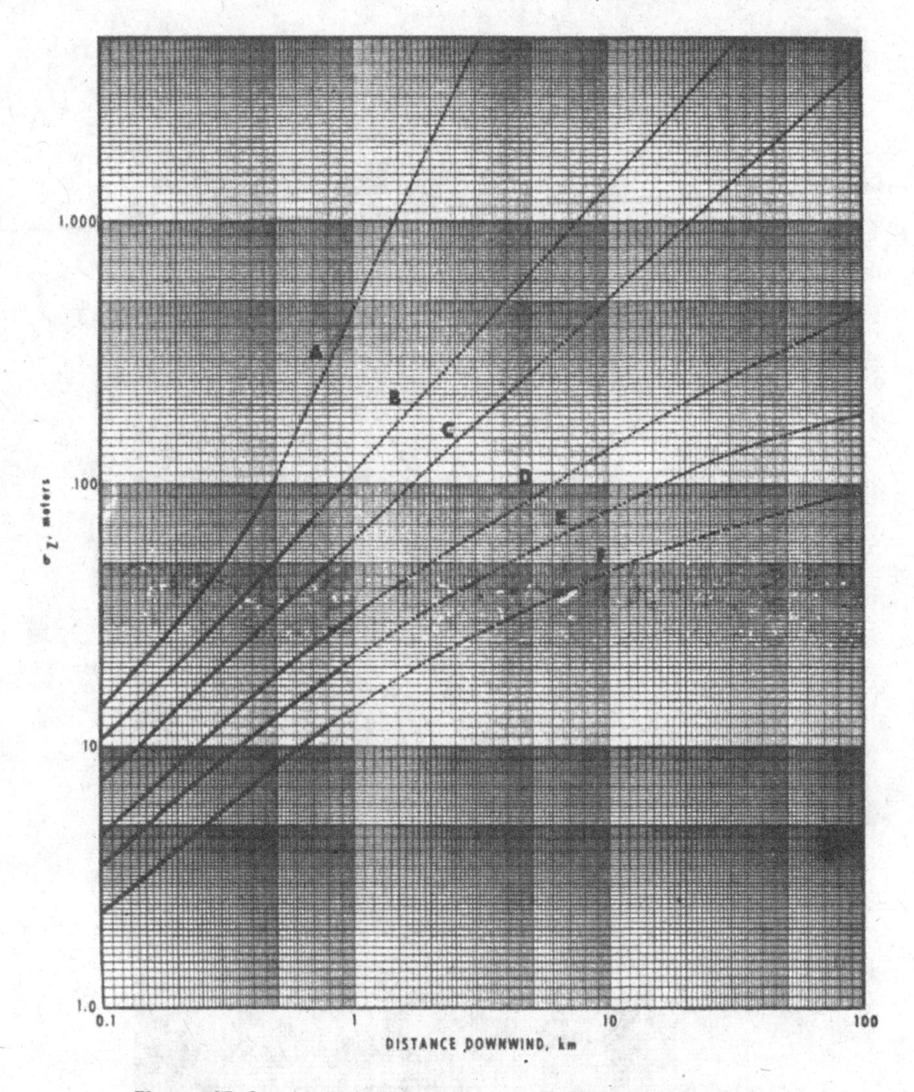

Figure 17–3 Vertical dispersion coefficient as a function of downwind distance from the source and stability category.

With this equation the concentration of the discharged aerosol can be determined at any point downwind of the source. Usually the maximum concentration of the discharged aerosol is of interest, at ground level. The maximum concentration will be directly downwind of the source, by inspection of Figure 17–1, coincident with the *X*-axis. Along the *X*-axis, $y = 0$. At ground level, $z = 0$. Inserting these values into Turner's equation, the equation for the maximum ground level concentration is as follows:

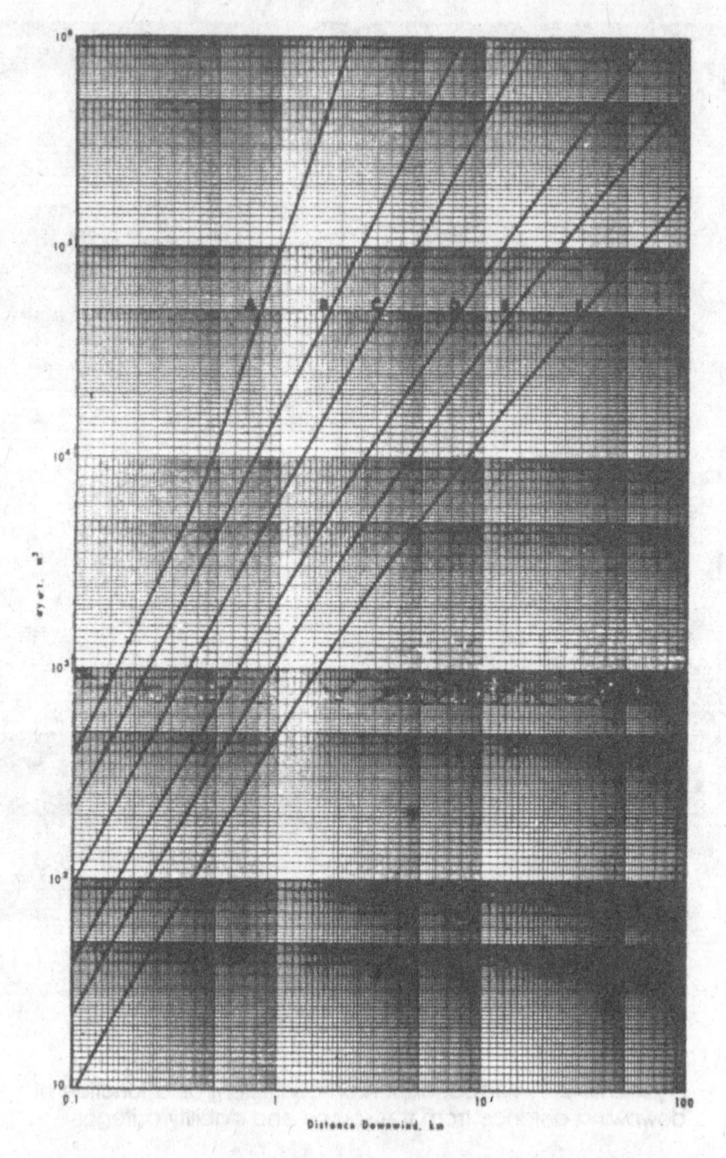

Figure 17–4 The product of $\sigma_y \sigma_z$ as a function of downwind distance from the source.

$$\omega\,(x,0,0,H) = \frac{Q}{\pi \sigma_y \sigma_z u} \cdot \exp\left\{-\frac{1}{2}\,(H/\sigma_z)^2\right\}$$

As noted previously, this equation provides a conservative approximation of the concentration of an aerosol remote from the source of the discharge.

Typical Example

Particulate matter is discharged from a stack at a rate of 8 pounds per hour. The stack exit is 60 feet above ground level and with a 3-foot outlet, the stack exit velocity is 50 feet per second. The ambient temperature is 80°F and the stack exit is 120°F. With a wind velocity of 8 miles per hour, what is the particulate concentration at ground level 2000 feet downstream of the stack? Assume that the atmospheric conditions are moderate daylight, strong wind, corresponding to Category D in Table 17–1. First, determine the effective stack height, from Holland's equation, as follows:

$$\Delta H = \frac{v \cdot d}{u}\left(1.5 + 0.00268 \cdot P \cdot \frac{T_s - T_a}{T_s} \cdot d\right)$$

With:
v = 50 ft/sec = 15.250 *m/s*
d = 3 ft = 0.914 *m*
u = 8 mph = 3.58 *m/s*
P = 14.7 psia = 213. *mb*
T_s = 120°F = 322K
T_a = 80° F = 300K

(Note that the above metric values were obtained from the conversion tables within the appendix.)

Substituting:

$$\Delta H = \frac{15.240 \cdot 0.914}{3.58}\left(1.5 + 0.00268 \cdot 213 \cdot \frac{322 - 300}{322} \cdot 0.914\right)$$

$$\Delta H = 6.0\ m$$

The value ΔH is the plume rise. The effective stack height is equal to the height of the stack (h = 60 ft = 18.3 m) plus the plume rise (ΔH = 6.0 m). In this case:

$$H = h + \Delta H = 18.3 + 6.0 = 24.3\ m, \text{ effective stack height.}$$

Now that the effective plume rise has been determined, the emission concentration can be calculated:

From Turner's equation:

$$\omega\ (x,0,0,H) = \frac{Q}{\pi \sigma_y \sigma_z u} \cdot \exp\left\{-\frac{1}{2}(H/\sigma_z)^2\right\}$$

where, from Figures 17–1 and 17–4, for 2000 feet (610 m = 0.61 km) and Category D:

$$\sigma_z = 22\ m$$
$$\sigma_y\sigma_z = 970\ m$$
and, from above,
$$H = 24.3\ m$$
$$Q = 8\ \text{lb/hr} = 1.01\ g/s$$

Substituting in Turner's equation:

$$\omega\ (x,0,0,H) = \frac{1.01}{970 \cdot 3.58\pi} \cdot \exp\left\{-\frac{1}{2}(24.3/22.0)^2\right\}$$

$$\omega = 0.000050\ g/m^3$$
with one $\mu g = 1/1000000\ g$,
$$\omega = 60\ \mu g/m^3$$

For this example, the particulate concentration is calculated as 50 micrograms per cubic meter 3000 feet from the source. Comparing this figure to the primary NAAQS for particulate matter (Table 3–2), which is 75 $\mu g/m^3$, the emission is safely within the acceptable range.

The above example was run for particulate matter. These equations are equally valid for the dispersion, or diffusion of other aerosols, gases, or even for the emission of radioactivity from a source. As noted previously, these calculations are conservative. The actual emissions as measured at the calculated point will normally be less than the Taylor approximation. In fact, other, more sophisticated techniques for calculating emissions (e.g., computer modeling) will usually produce lesser emission concentration values than the above equations.

References and Bibliography

1. Turner, D. 1969. *Workbook of atmospheric dispersion estimates.* Cincinnati, OH: US Department of Health, Education and Welfare, Public Health Service, National Air Pollution Control Administration.
2. Holland, J. 1953. *A Meteorological survey of the Oak Ridge area.* Washington, DC: US Atomic Energy Commission Report ORO-99.
3. California State Air Resource Board. 1974. *Introduction to manual methods for estimating air quality.* California State Air Resources Board, Washington, DC: Government Printing Office, July.
4. Briggs, G. 1973. *Diffusion estimation for small emissions,* EPA, Washington, DC: Government Printing Office, May.
5. Barnes, H. 1981. *Characterization of scrubbed and unscrubbed power plant plumes.* EPA 600/S3-81-041. Washington, DC: Government Printing Office, October.

18

Noise Generation and Control

Noise is an air discharge which can cause immediate public concern and reaction. It is at times the most noticeable of discharges from an industrial system and can often be difficult and expensive to control.

Noise

Noise is generally regarded as any sound that may produce an undesired physiological or psychological effect in an individual or animal or that may interfere with the social end of an individual or group. Those ends include all human activities: communication, work, rest, recreation, and sleep. Also, any sound having an adverse effect on property and/or material is considered "noise."

The above definition is not absolute. Noise is not differentiated as to intensity, duration, pattern, or frequency. Noise is regarded as such by purely subjective means. A sound is "noise" when it creates a negative impact on a receptor, whether a human being, animal, or inanimate object. A corollary of this definition is that anyone may declare that noise exists, and the burden of proof is likely to be on the generator, not necessarily on the observer. Often, particularly when a generator is a public facility, or an industrial facility, the generator must demonstrate to the public that a nuisance is not being created; for example, that the facility is not generating noise, or objectionable sounds.

Noise Measurement

Sound is measured against two basic scales: sound power level and sound pressure level.

Sound power is measured in watts of electricity radiated by a source of sound. Most noise in the environment originates from sources radiating less than a single watt of power. As a reference point however, a figure of one-trillionth of a watt (10^{-12} watt) is used as a base against which sound power level is measured. The lowest audible sound, a soft whisper, radiates approximately one-billionth (10^{-9}).

The sound pressure level is most commonly used for the measurement and description of noise. It is a measure, in terms of atmospheric pressure, of sound as it passes a point any given distance from a sound source. The most common reference point for the sound pressure level is 0.0002 of 1 microbar, where a bar is defined as atmospheric pressure at sea level, 14.7 psia at a temperature of 59°F. The unit Pascal (Pa) is also in common usage in measurement of sound. In terms of the common reference point for sound pressure level, 20μPa is equivalent to 0.0002 microbars.

Decibels are used to relate the intensity of one sound to that of another. A decibel is a logarithmic expression, defined as follows:

$$\text{Sound Power Level} = 10 \log (\text{Watts}/10^{-12}) \text{ d}\beta$$

$$\text{Sound Pressure Level} = 20 \log (\text{microbars}/0.0002) \text{ d}\beta$$

If the detected pressure (microbars) of a noise is, for instance, 1000 times higher than the base sound level, the noise will have a sound pressure level of 20 log (1000) = 60 dβ.

Of greatest interest is the response of the human ear to a particular noise. The ear does not respond equally to all frequencies, but is less efficient at low and high frequencies than it is at medium or speech range frequencies. Thus, to obtain a single number representing the sound level of a noise containing a wide range of frequencies in a manner representative of the ear's response, it is necessary to reduce, or weight the effects of the low or high frequencies with respect to the medium frequencies. The resultant sound level is said to be A-weighted and the results are expressed as dBA. Most sound level meters have an internal A-weighting network for directly measuring the A-weighted sound level.

The A-weighted sound level system is illustrated in Figure 18–1, which also describes other weighting standards. The IEC Standard refers to Recommendations 123 and 179 of the International Electrotechnical Commission.

Human Response to Noise

The response of individuals to noise includes annoyance and health hazards. Hearing loss can occur from two major sources (besides accidents and congenital effects):

- Presbycusis, a decline in hearing acuity due to old age.
- Long duration exposure to high levels of noise.

The basic hazardous noise criteria that have been used for the workplace are that a noise environment is unacceptable if after 10 years of 8-hour per day exposure, the average employee has suffered a permanent work-induced hearing loss of:

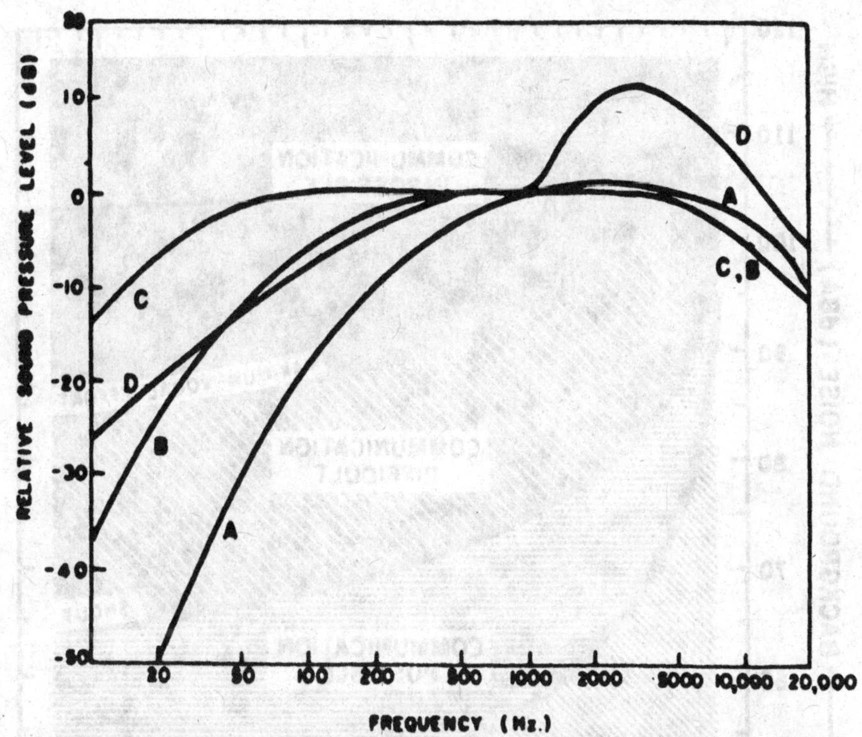

Also shown is the proposed D weighting curve for monitoring jet aircraft noise. From the curves it can be seen that for a 50 Hz pure tone the reading on the A scale (which discriminates against low frequency sounds) would be 30 dB less than the C scale reading.

Figure 18–1 IEC standard A, B, C, weighting curves for standard sound level meters.

Table 18–1 OSHA Noise Exposure Levels

Duration hr/day	dBA Response	
	Maximum Sound Level	Recommended Sound Level
8	85	75
4	88	78
2	91	81
1	94	84
0.50	97	87
0.25 or less	100	90

Note: Personnel protective equipment should be supplied and used whenever reduction to the recommended levels in this table is not feasible.

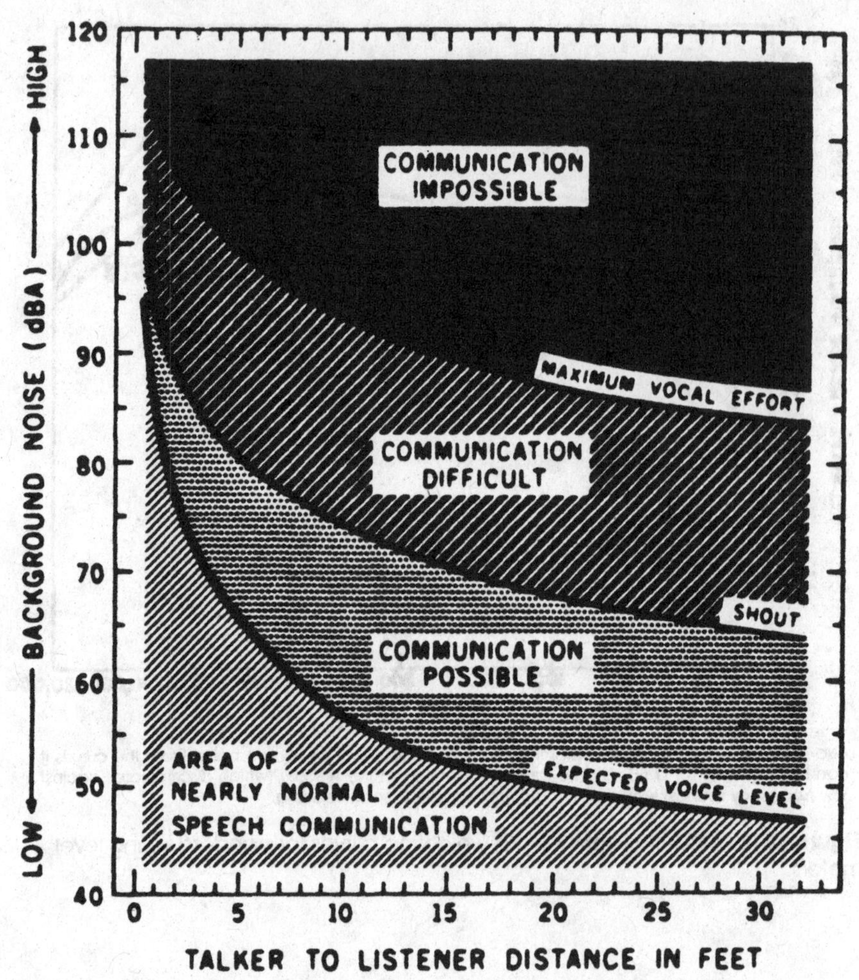

Figure 18–2 Simplified diagram showing speech communications as a function of background noise and talker-listener distance. (*Source: M. Kerbec, "Noise & Hearing," Output Systems Corp. (1972): 7–35.*)

- 10 dB at 1000 Hz or
- 15 dB at 2000 Hz or
- 20 dB at 3000 Hz or above.

In accordance with these criteria the U.S. Occupational Health and Safety Administration (OSHA), has established maximim allowable noise levels for the workplace, as listed in Table 18–1. Figure 18–2 illustrates the relationship between noise level and distance between a talker and a listener. This is an indicator of acceptable and unacceptable noise in an everyday environment.

Noise Generation

Noises generated within an incineration facility are normally those resulting from the movement of air or gas. Fans and blowers will create the greatest noise levels and of secondary concern is the noise generated through ducts, flues, and nozzles. In the combustion process, the burning of a fuel within a fixed chamber, noise will also be generated.

There are two components resulting from a generated noise: local noise, which may have a significant impact on the health and safety of operating personnel, and noise broadcast from the facility to the surrounding area, which as the potential to create public issues and public responses.

Typical interior and exterior noise levels are listed in Table 18–2. Facility noise should be compared to existing noise level as noted in this table.

Controlling Incineration Noise Generation

There are many components of incineration systems which are inherent noise sources. There are means for controlling some of this generated noise:

Flame Noise

The flame envelope within the furnace constantly changes as fuel-rich pockets of gases or vapors meet with oxygen and burn in "puffs." Similarly, as oxygen-rich pockets mix with fuel, they enter the flammability range, and "puff." The resulting noise, caused by the moving flame fronts, and related to the rate of change in flame surface area, is termed *combustion roar*. Combustion-driven oscillations result when a furnace feed-back mechanism produces oscillations that lead to a variable heat release, the timing of which is such that it adds energy to subsequent oscillations. As an example, a pulsating blower not only generates noise, but provides a mechanism for combustion driven oscillations.

Combustion-driven oscillations involve a feedback cycle that converts chemical or thermal energy to oscillatory energy in the fluid stream. This is an unusually efficient conversion that can yield a near pure tone at a frequency less than 50 Hz.

Superturbulent noise is a special form of combustion roar associated with unstable flames and evidenced by a sudden jump in the sound pressure level. This results from the increased turbulence in an unstable operating mode, such as occurs when a flame fluctuates between tiles, when a flame detaches from a burner tile, or when a burner is fired beyond its rated range. Such instability in a flame front can double the efficiency of conversion of chemical and thermal energy to noise.

Table 18–2 Average Single Number Sound Levels, dBA

Interior Noises	
Bedroom at night	30–40
Quiet residence	39–48
Residence with radio	47–59
Small office or store	47–59
Large store	52–63
Large office	57–68
Electric typewriter at 10 feet	62–67
Factory office	60–73
Automobile	64–78
Factory	65–93
School cafeteria	76–85
Railroad car	77–88
Garbage disposal	78–83
Airplane cabin	88–98
Noises 3 Feet From Source	
Whispering	30–35
Quiet ventilation outlet	41–47
Quiet talking	59–66
Noisy ventilation outlet	60–75
Business machine	71–86
Lathe	73–83
Shouting	74–80
Power saw	93–101
Power mower	94–102
Farm tractor	94–103
Power wood planer	97–108
Pneumatic riveter	110–120
Outside Noises	
Leaves rustling	10–15
Bird call	40–45
Quiet residential street	40–52
150 feet to 200 feet from dense traffic	55–70
Edge of highway with dense traffic	70–85
Car at 65 mph at 25 feet	75–80
Propellor plane at 100 feet	75–84
Pneumatic drill at 50 feet	80–85
Noisy street	84–94
Under elevated train	88–97
Jet plane at 1000 feet	100–105
Jet take-off at 200 feet	120–125
50-hp siren at 100 feet	130–135

Burner noise level tends to increase with greater flow velocity, flame velocity, swirl, firing rate, port diameter, and length. As industry converts to low-BTU gas worse noise problems may appear precipitated by gas fuel's higher flame velocities.

For a given nozzle configuration doubling the firing rate usually increase the noise level by about 6 dBA. A 3 dBA reduction may be possible by lowering the noise level by 35%. Good (low) noise conditions and good (low) NO_x pollutant emissions usually go together but, unfortunately, they are also accompanied by poorer (lower) heat release and lower flame velocity resulting in less uniform heating with consequent fuel waste.

Flames often make less noise if burning off ratio; that is, too rich or too lean to be most efficient. Another variable is swirl, which affects mixing rate. Variable heat pattern burners, which have a external means for adjustment of register position (such as a push rod), offer the opportunity for experimentation. The objective of this feature, however, is not noise control but to control flame shape and heat transfer.

One way to minimize combustion roar is to use more small burners, in place of a few large ones. Similarly, use more burners with less pressure drop; that is, less capacity per burner, but the same total capacity within the furnace.

Eliminate furnace openings. Use sealed-in burners instead of open burners with registers or shutters. This also saves fuel by assuring control of fuel/air ration at all firing rates. Better scheduling, to minimize the time that charge and unload doors are open, also reduces supplemental fuel consumption and the daily noise dose of workers.

Blower System Noises

Blower system noises originate from blower inrush noise, impeller tip speed noise, pulsation, casing vibration, and motor, turbine, or engine

Table 18–3 Typical Values of Peak Sound Pressure Levels For Impulse Noise

SPL	Example
190 +	Within blast zone of exploding bomb.
160–180	Within crew area of heavy artillery piece or naval gun when shooting.
140–170	At shooter's ear when firing handgun.
125–160	At child's ear when detonating toy cap or firecracker.
120–140	Metal to metal impacts in many industrial processes (e.g., drop forging, metal beating) adjacent to listener.
110–130	On construction site during pile driving adjacent to listener.

Note: All SPLs in dB re 20μPa.

drive noise. To reduce this noise reduce tip speed, and, wherever possible, select low rpm, multistage blowers with smaller impeller diameters.

The eye or inlet of a blower is a form of orifice where turbulent flow over the inlet lip can generate sound. A protective inlet grill may generate noise similar to the action of a vibrating reed. Any loose flanged inlet accessory (valve, filter, or silencer) may leave a slit or a gap which can act as a whistle. A piped intake or a blower discharge pipe may be of such a length that it can produce an organ-pipe effect. Locating a blower remote from a furnace not only removes one noise source but may also provide a cleaner air source, away from the severe ash and dust-laden atmosphere in the immediate area of the incinerator.

Flow Noise

Flow noise is generated by the motion of air and fuel through a burner, piping, valves, and accessories. To minimize the noise level from this source design for low velocities and few flow restrictions. It is easy to check for burner air nozzle noise by shutting off the supply of supplemental fuel. If the noise continues after the flame goes out, air nozzle noise was the cause. It is dangerous to attempt to check for burner fuel nozzle noise by turning off the air unless the chamber is cold (pilots off) and its atmosphere can be thoroughly vented to the outdoors (not the room) during and for a long time after the test. Fuel nozzle noise, however, is a less common problem, except in the case of inspirators. (Inspirators can be horrendous noise emitters because they often have such a large high-pressure drop across the spud that sonic velocity results. The large venturi often used with inspirators in refineries can frequently act as a sounding board.)

Burner nozzle noise usually has higher frequency than that of flame noise. Using larger connections and feed pipes sometimes helps decrease noise level. Noise generated at such restrictions is a steep logarithmic function of the flow velocity. Selecting burners with lower flow velocities (largest possible fuel jets for the needed flame shape) minimizes both flow noise and flame noise.

Valves and Piping Noises

Aerodynamic noise from valves and orifices can be minimized by selection of designs with minimim turbulence, low velocities, and no vibrating parts, and by damping the transmission of the noise through the associated piping, as by a muffler section (oversized and lined with acoustic insulation). Sound insulating materials should be installed around components that cannot be muffled internally.

Duct or piping noise is more likely to be a problem if flow velocities are high or if there are frequent turbulence inducers such as sudden expansions or flow obstructions. There is less chance that inside noise will be transmitted through rigid duct walls. However, flexible ducting reduces the transmission of vibrations along the length of a duct. Both vibration and noise transmission can be reduced by use of in-line silencers and acoustical installation.

Interactions

If the sound frequency emitted by any of the above sources happens to coincide with the natural frequency of the room, furnace enclosure, piping or ducting, breeching or stack, or the housings of blowers or auxiliaries, harmonic excitations will amplify the sound. If they do not coincide, there may be an attenuating or muffling effect.

Flames have been known to amplify the high-frequency flow noise of a supply line by as much as 20dB. A flame provides the energy to excite standing waves, and if their frequency matches the resonant frequency of the surrounding furnace or room, a loud, low-pitched noise will result. It is difficult to determine where a noise originates, particularly if it is being amplified by another element in the system, such as duct excitation or valve vibration.

Suppression of combustion noise amplification by room or furnace acoustics is best accomplished by altering the position of the combustion noise source relative to the walls so as to change the natural modes of the standing waves. Although openings in a furnace wall around a burner may alleviate standing waves, there is usually a better chance of a net overall noise reduction through the muffling effect of a well-sealed furnace.

Case Study

A 1500 hp motor drives an Induced Draft fan at a hazardous waste incineration facility in Cincinnati, Ohio. The fan turns at 1200 rpm, or 20 revolutions per second. It has eight blades and at 20 rps the beat frequency is 20 x 8 or 160 Hz.

Residents of an apartment complex almost 1/4 mile from the facility complained of a low droning noise. This noise was identified as occurring when the incinerations facility was in operation. Upon further investigation, the noise was found to be broadcast from the incinerator stack, with the prime frequency peak at 160 Hz and lesser peaks in integral multiples of 160; that is, 320 Hz, 480 Hz, etc. The noise appeared to be broadcast above the immediate surroundings, where it was not detected, to the apartment complex, some distance away, as illustrated in Figure 18–3.

In an attempt to reduce the noise at the receptor (the apartment complex), an insertion silencer was designed to absorb acoustical energy within the stack, above the induced draft fan. It was of the splitter type with perforated sections filled with sound absorbing mineral wool, and

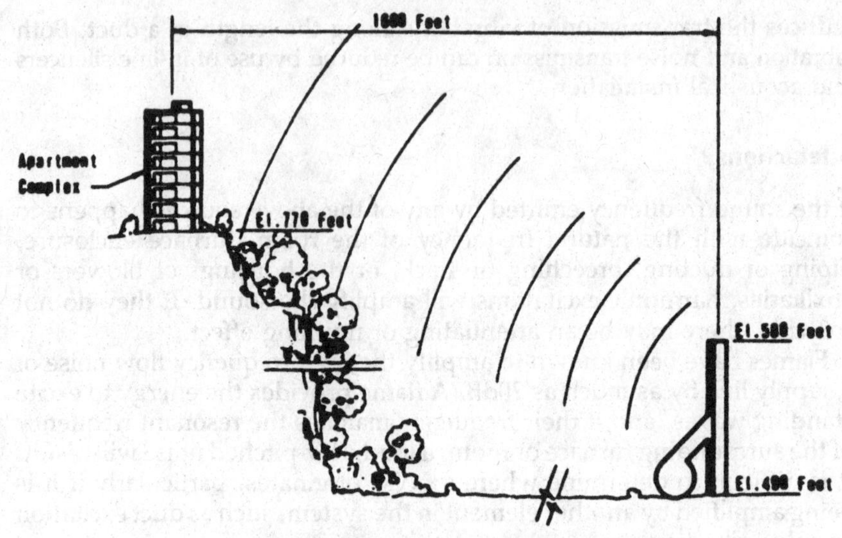

Figure 18–3 Broadcast noise (not to scale).

Table 18–4 Sound Pressure Levels in Octave Bands at the Apartments With ID Fan on at Full Capacity

	63	125	250	500	1000	2000	4000	A-Level
Before Silencer	72	78	70	64	59	54	43	68
After Silencer	69	67	60	57	52	46	37	59
Background	71	67	62	58	56	50	43	61

provisions were included to prevent closing of the perforations by a buildup of particulate from the flowing exhaust gas stream.

Resonating type silencers were considered for this application but they were rejected in favor of the noise absorbing design. Although the noise generated was pure tone (160 Hz), the complexity of the equipment arrangement would not assure satisfactory attenuation if the pure tone were eliminated from broadcast by resonance. This tone, although eliminated from broadcast, was still a generated tone, and could have created other tonal frequencies which would find a resonator transparent.

This silencer proved to be extremely effective. As shown in Table 18–4, the broadcast noise was reduced below that of the background. (Measurements of background noise and noise after the silencer was installed were made at different dates. This fact accounts for those figures in this table which indicate that the silencer reduced background noise, a physical impossibility.) The single tone reduction at 160 Hz was greater than 20 dB.

References and Bibliography

1. Magan, A. Quiet communities: Minimizing the effects of noise through land use controls. National Association of Counties Research, Inc., March 1979.
2. Noise and the environment. *USEPA Journal* (October 1979) 5/9.
3. Liebich, R., and P., Ostergaard. Industrial noise pollution. *Mechanical Engineering* (July 1981).
4. Weihsmann, P. Noise management by reverberation control. *Pollution Engineering* (February 1978): 56–57.
5. American Industrial Hygiene Association. 1975. *Industrial Noise Manual*. Westmont, NJ: AIHA.
6. Kinsley, G. Specifying sound levels for new equipment. *Chemical Engineering* (June 18, 1979).
7. Brunner, C., and J. Trapp. Progress report of industrial liquid fluid thermal processing system. *Proceedings of the American Society of Mechanical Engineers, Solid Waste Processing Division* Washington, D.C., May 1980.

Appendix

Table A-1

CONVERSION --- METERS VS. INCHES

METERS		INCHES	METERS		INCHES	METERS		INCHES
0.000	0.01	0.39	0.069	2.7	106.30	0.432	17.	669.29
0.001	0.02	0.79	0.071	2.8	110.24	0.457	18.	708.66
0.001	0.03	1.18	0.074	2.9	114.17	0.483	19.	748.03
0.001	0.04	1.57	0.076	3.0	118.11	0.508	20.	787.40
0.001	0.05	1.97	0.079	3.1	122.05	0.533	21.	826.77
0.002	0.06	2.36	0.081	3.2	125.98	0.559	22.	866.14
0.002	0.07	2.76	0.084	3.3	129.92	0.584	23.	905.51
0.002	0.08	3.15	0.086	3.4	133.86	0.601	24.	944.88
0.002	0.09	3.54	0.089	3.5	137.80	0.635	25.	984.25
0.003	0.10	3.94	0.091	3.6	141.73	0.660	26.	1023.60
0.005	0.2	7.87	0.094	3.7	145.67	0.686	27.	1062.99
0.008	0.3	11.81	0.097	3.8	149.61	0.711	28.	1102.36
0.010	0.4	15.75	0.099	3.9	153.54	0.737	29.	1141.73
0.013	0.5	19.69	0.102	4.0	157.48	0.762	30.	1181.10
0.015	0.6	23.62	0.104	4.1	161.42	0.787	31.	1220.47
0.018	0.7	27.56	0.107	4.2	165.35	0.813	32.	1259.84
0.020	0.8	31.50	0.109	4.3	169.29	0.838	33.	1299.21
0.023	0.9	35.43	0.112	4.4	173.23	0.864	34.	1338.58
0.025	1.0	39.37	0.114	4.5	177.17	0.889	35.	1377.95
0.028	1.1	43.31	0.117	4.6	181.10	0.914	36.	1417.32
0.030	1.2	47.24	0.119	4.7	185.04	0.940	37.	1456.69
0.033	1.3	51.18	0.122	4.8	188.98	0.965	38.	1496.06
0.036	1.4	55.12	0.124	4.9	192.91	0.991	39.	1535.43
0.038	1.5	59.06	0.127	5.0	196.85	1.016	40.	1574.80
0.041	1.6	62.99	0.152	6.	236.22	1.041	41.	1614.17
0.043	1.7	66.93	0.178	7.	275.59	1.067	42.	1653.54
0.046	1.8	70.87	0.203	8.	314.96	1.092	43.	1692.91
0.048	1.9	74.80	0.229	9.	354.33	1.118	44.	1732.28
0.051	2.0	78.74	0.254	10.	393.70	1.143	45.	1771.65
0.053	2.1	82.68	0.279	11.	433.07	1.168	46.	1811.02
0.056	2.2	86.61	0.305	12.	472.44	1.194	47.	1850.39
0.058	2.3	90.55	0.330	13.	511.81	1.219	48.	1889.76
0.061	2.4	94.49	0.356	14.	551.18	1.245	49.	1929.13
0.064	2.5	98.43	0.381	15.	590.55	1.270	50.	1968.50
0.066	2.6	102.36	0.406	16.	629.92			

Example: 2 meters = 78.74 inches
4.8 inches = 0.122 meters

Table A-2

CONVERSION --- METERS VS. FEET

METERS		FEET	METERS		FEET	METERS		FEET
0.003	0.01	0.03	1.311	4.3	14.11	2.865	9.4	30.84
0.006	0.02	0.07	1.341	4.4	14.44	2.896	9.5	31.17
0.009	0.03	0.10	1.372	4.5	14.76	2.926	9.6	31.50
0.012	0.04	0.13	1.402	4.6	15.09	2.957	9.7	31.82
0.015	0.05	0.16	1.433	4.7	15.42	2.987	9.8	32.15
0.018	0.06	0.20	1.463	4.8	15.75	3.018	9.9	32.48
0.021	0.07	0.23	1.494	4.9	16.08	3.048	10.0	32.81
0.024	0.08	0.26	1.524	5.0	16.40	3.078	10.1	33.14
0.027	0.09	0.30	1.554	5.1	16.73	3.109	10.2	33.46
0.030	0.10	0.33	1.585	5.2	17.06	3.139	10.3	33.79
0.061	0.2	0.66	1.615	5.3	17.39	3.170	10.4	34.12
0.091	0.3	0.98	1.646	5.4	17.72	3.200	10.5	34.45
0.122	0.4	1.31	1.676	5.5	18.04	3.231	10.6	34.78
0.152	0.5	1.64	1.707	5.6	18.37	3.261	10.7	35.10
0.183	0.6	1.97	1.737	5.7	18.70	3.292	10.8	35.43
0.213	0.7	2.30	1.768	5.8	19.03	3.322	10.9	35.76
0.244	0.8	2.62	1.798	5.9	19.36	3.353	11.0	36.09
0.274	0.9	2.95	1.829	6.0	19.69	3.383	11.1	36.42
0.305	1.0	3.28	1.859	6.1	20.01	3.414	11.2	36.75
0.335	1.1	3.61	1.890	6.2	20.34	3.444	11.3	37.07
0.366	1.2	3.94	1.920	6.3	20.67	3.475	11.4	37.40
0.396	1.3	4.27	1.951	6.4	21.00	3.505	11.5	37.73
0.427	1.4	4.59	1.981	6.5	21.33	3.536	11.6	38.06
0.457	1.5	4.92	2.012	6.6	21.65	3.566	11.7	38.39
0.488	1.6	5.25	2.042	6.7	21.98	3.597	11.8	38.71
0.518	1.7	5.58	2.073	6.8	22.31	3.627	11.9	39.04
0.549	1.8	5.91	2.103	6.9	22.64	3.658	12.0	39.37
0.579	1.9	6.23	2.134	7.0	22.97	3.688	12.1	39.70
0.610	2.0	6.56	2.164	7.1	23.29	3.719	12.2	40.03
0.640	2.1	6.89	2.195	7.2	23.62	3.749	12.3	40.35
0.671	2.2	7.22	2.225	7.3	23.95	3.780	12.4	40.68
0.701	2.3	7.55	2.256	7.4	24.28	3.810	12.5	41.01
0.732	2.4	7.87	2.286	7.5	24.61	3.840	12.6	41.34
0.762	2.5	8.20	2.316	7.6	24.93	3.871	12.7	41.67
0.792	2.6	8.53	2.347	7.7	25.26	3.901	12.8	41.99
0.823	2.7	8.86	2.377	7.8	25.59	3.932	12.9	42.32
0.853	2.8	9.19	2.408	7.9	25.92	3.962	13.0	42.65
0.884	2.9	9.51	2.438	8.0	26.25	3.993	13.1	42.98
0.914	3.0	9.84	2.469	8.1	26.57	4.023	13.2	43.31
0.945	3.1	10.17	2.499	8.2	26.90	4.054	13.3	43.64
0.975	3.2	10.50	2.530	8.3	27.23	4.084	13.4	43.96
1.006	3.3	10.83	2.560	8.4	27.56	4.115	13.5	44.29
1.036	3.4	11.15	2.591	8.5	27.89	4.145	13.6	44.62
1.067	3.5	11.48	2.621	8.6	28.22	4.176	13.7	44.95
1.097	3.6	11.81	2.652	8.7	28.54	4.206	13.8	45.28
1.128	3.7	12.14	2.682	8.8	28.87	4.237	13.9	45.60
1.158	3.8	12.47	2.713	8.9	29.20	4.267	14.0	45.93
1.189	3.9	12.80	2.743	9.0	29.53	4.298	14.1	46.26
1.219	4.0	13.12	2.774	9.1	29.86	4.328	14.2	46.59
1.250	4.1	13.45	2.804	9.2	30.18	4.359	14.3	46.92
1.280	4.2	13.78	2.835	9.3	30.51	4.389	14.4	47.24

Table A-2

(continued)

METERS		FEET	METERS		FEET	METERS		FEET
4.420	14.5	47.57	6.005	19.7	64.63	21.031	69.	226.38
4.450	14.6	47.90	6.035	19.8	64.96	21.336	70.	229.66
4.481	14.7	48.23	6.066	19.9	65.29	21.641	71.	232.94
4.511	14.8	48.56	6.096	20.0	65.62	21.946	72.	236.22
4.542	14.9	48.88	6.401	21.	68.90	22.250	73.	239.50
4.572	15.0	49.21	6.706	22.	72.18	22.555	74.	242.78
4.602	15.1	49.54	7.010	23.	75.46	22.860	75.	246.06
4.633	15.2	49.87	7.315	24.	78.74	23.165	76.	249.34
4.663	15.3	50.20	7.620	25.	82.02	23.470	77.	252.62
4.694	15.4	50.52	7.925	26.	85.30	23.774	78.	255.91
4.724	15.5	50.85	8.230	27.	88.58	24.079	79.	259.19
4.755	15.6	51.18	8.534	28.	91.86	24.384	80.	262.47
4.785	15.7	51.51	8.839	29.	95.14	24.689	81.	265.75
4.816	15.8	51.84	9.144	30.	98.43	24.994	82.	269.03
4.846	15.9	52.17	9.449	31.	101.71	25.298	83.	272.31
4.877	16.0	52.49	9.754	32.	104.99	25.603	84.	275.59
4.907	16.1	52.82	10.058	33.	108.27	25.908	85.	278.87
4.938	16.2	53.15	10.363	34.	111.55	26.213	86.	282.15
4.968	16.3	53.48	10.668	35.	114.83	26.518	87.	285.43
4.999	16.4	53.81	10.973	36.	118.11	26.822	88.	288.71
5.029	16.5	54.13	11.278	37.	121.39	27.127	89.	291.99
5.060	16.6	54.46	11.582	38.	124.67	27.432	90.	295.28
5.090	16.7	54.79	11.887	39.	127.95	27.737	91.	298.56
5.121	16.8	55.12	12.192	40.	131.23	28.042	92.	301.84
5.151	16.9	55.45	12.497	41.	134.51	28.346	93.	305.12
5.182	17.0	55.77	12.802	42.	137.80	28.651	94.	308.40
5.212	17.1	56.10	13.106	43.	141.08	28.956	95.	311.68
5.243	17.2	56.43	13.411	44.	144.36	29.261	96.	314.96
5.273	17.3	56.76	13.716	45.	147.64	29.566	97.	318.24
5.304	17.4	57.09	14.021	46.	150.92	29.870	98.	321.52
5.334	17.5	57.41	14.326	47.	154.20	30.175	99.	324.80
5.364	17.6	57.74	14.630	48.	157.48	30.480	100.	328.08
5.395	17.7	58.07	14.935	49.	160.76	34.	110.	361.
5.425	17.8	58.40	15.240	50.	164.04	37.	120.	394.
5.456	17.9	58.73	15.545	51.	167.32	40.	130.	427.
5.486	18.0	59.06	15.850	52.	170.60	43.	140.	459.
5.517	18.1	59.38	16.154	53.	173.88	46.	150.	492.
5.547	18.2	59.71	16.459	54.	177.17	49.	160.	525.
5.578	18.3	60.04	16.764	55.	180.45	52.	170.	558.
5.608	18.4	60.37	17.069	56.	183.73	55.	180.	591.
5.639	18.5	60.70	17.374	57.	187.01	58.	190.	623.
5.669	18.6	61.02	17.678	58.	190.29	61.	200.	656.
5.700	18.7	61.35	17.983	59.	193.57	64.	210.	689.
5.730	18.8	61.68	18.288	60.	196.85	67.	220.	722.
5.761	18.9	62.01	18.593	61.	200.13	70.	230.	755.
5.791	19.0	62.34	18.898	62.	203.41	73.	240.	787.
5.822	19.1	62.66	19.202	63.	206.69	76.	250.	820.
5.852	19.2	62.99	19.507	64.	209.97	79.	260.	853.
5.883	19.3	63.32	19.812	65.	213.25	82.	270.	886.
5.913	19.4	63.65	20.117	66.	216.54	85.	280.	919.
5.944	19.5	63.98	20.422	67.	219.82	88.	290.	951.
5.974	19.6	64.30	20.726	68.	223.10	91.	300.	984.

Table A-2

(continued)

METERS		FEET	METERS		FEET	METERS		FEET
94.	310.	1017.	213.	700.	2297.	579.	1900.	6234.
98.	320.	1050.	216.	710.	2329.	610.	2000.	6562.
101.	330.	1083.	219.	720.	2362.	640.	2100.	6890.
104.	340.	1115.	223.	730.	2395.	671.	2200.	7218.
107.	350.	1148.	226.	740.	2428.	701.	2300.	7546.
110.	360.	1181.	229.	750.	2461.	732.	2400.	7874.
113.	370.	1214.	232.	760.	2493.	762.	2500.	8202.
116.	380.	1247.	235.	770.	2526.	792.	2600.	8530.
119.	390.	1280.	238.	780.	2559.	823.	2700.	8858.
122.	400.	1312.	241.	790.	2592.	853.	2800.	9186.
125.	410.	1345.	244.	800.	2625.	884.	2900.	9514.
128.	420.	1378.	247.	810.	2657.	914.	3000.	9843.
131.	430.	1411.	250.	820.	2690.	945.	3100.	10171.
134.	440.	1444.	253.	830.	2723.	975.	3200.	10499.
137.	450.	1476.	256.	840.	2756.	1006.	3300.	10827.
140.	460.	1509.	259.	850.	2789.	1036.	3400.	11155.
143.	470.	1542.	262.	860.	2822.	1067.	3500.	11483.
146.	480.	1575.	265.	870.	2854.	1097.	3600.	11811.
149.	490.	1608.	268.	880.	2887.	1128.	3700.	12139.
152.	500.	1640.	271.	890.	2920.	1158.	3800.	12467.
155.	510.	1673.	274.	900.	2953.	1189.	3900.	12795.
158.	520.	1706.	277.	910.	2986.	1219.	4000.	13123.
162.	530.	1739.	280.	920.	3018.	1524.	5000.	16404.
165.	540.	1772.	283.	930.	3051.	1829.	6000.	19685.
168.	550.	1804.	287.	940.	3084.	2134.	7000.	22966.
171.	560.	1837.	290.	950.	3117.	2438.	8000.	26247.
174.	570.	1870.	293.	960.	3150.	2743.	9000.	29258.
177.	580.	1903.	296.	970.	3182.	3048.	10000.	32808.
180.	590.	1936.	299.	980.	3215.	3353.	11000.	36089.
183.	600.	1969.	302.	990.	3248.	3658.	12000.	39370.
186.	610.	2001.	305.	1000.	3281.	3962.	13000.	42651.
189.	620.	2034.	335.	1100.	3609.	4267.	14000.	45932.
192.	630.	2067.	366.	1200.	3937.	4572.	15000.	49213.
195.	640.	2100.	396.	1300.	4265.	4877.	16000.	52493.
198.	650.	2133.	427.	1400.	4593.	5182.	17000.	55774.
201.	660.	2165.	457.	1500.	4821.	5486.	18000.	59055.
204.	670.	2198.	488.	1600.	5249.	5791.	19000.	62336.
207.	680.	2231.	518.	1700.	5577.	6096.	20000.	65617.
210.	690.	2264.	549.	1800.	5906.			

Example: 1.2 meters = 3.94 feet
81 feet = 24.689 meters

Table A-3

CONVERSION ------------------ MILES VS. KILOMETERS

MILES	KILOMETERS		MILES	KILOMETERS		MILES	KILOMETERS	
0.06	0.1	0.16	3.29	5.3	8.53	9.32	15.	24.14
0.12	0.2	0.32	3.36	5.4	8.69	9.94	16.	25.75
0.19	0.3	0.48	3.42	5.5	8.85	10.56	17.	27.36
0.25	0.4	0.64	3.48	5.6	9.01	11.19	18.	28.97
0.31	0.5	0.80	3.54	5.7	9.17	11.81	19.	30.58
0.37	0.6	0.97	3.60	5.8	9.33	12.43	20.	32.19
0.43	0.7	1.13	3.67	5.9	9.49	13.05	21.	33.79
0.50	0.8	1.29	3.73	6.0	9.66	13.67	22.	35.40
0.56	0.9	1.45	3.79	6.1	9.82	14.29	23.	37.01
0.62	1.0	1.61	3.85	6.2	9.98	14.91	24.	38.62
0.68	1.1	1.77	3.91	6.3	10.14	15.54	25.	40.23
0.75	1.2	1.93	3.98	6.4	10.30	16.16	26.	41.84
0.81	1.3	2.09	4.04	6.5	10.46	16.78	27.	43.45
0.87	1.4	2.25	4.10	6.6	10.62	17.40	28.	45.06
0.93	1.5	2.41	4.16	6.7	10.78	18.02	29.	46.67
0.99	1.6	2.57	4.23	6.8	10.94	18.64	30.	48.28
1.06	1.7	2.74	4.29	6.9	11.10	19.26	31.	49.89
1.12	1.8	2.90	4.35	7.0	11.26	19.88	32.	51.50
1.18	1.9	3.06	4.41	7.1	11.43	20.51	33.	53.11
1.24	2.0	3.22	4.47	7.2	11.59	21.13	34.	54.72
1.30	2.1	3.38	4.54	7.3	11.75	21.75	35.	56.32
1.37	2.2	3.54	4.60	7.4	11.91	22.37	36.	57.93
1.43	2.3	3.70	4.66	7.5	12.07	22.99	37.	59.54
1.49	2.4	3.86	4.72	7.6	12.23	23.61	38.	61.15
1.55	2.5	4.02	4.78	7.7	12.39	24.23	39.	62.76
1.62	2.6	4.18	4.85	7.8	12.55	24.86	40.	64.37
1.68	2.7	4.35	4.91	7.9	12.71	25.48	41.	65.98
1.74	2.8	4.51	4.97	8.0	12.87	26.10	42.	67.59
1.80	2.9	4.67	5.03	8.1	13.04	26.72	43.	69.20
1.86	3.0	4.83	5.10	8.2	13.20	27.34	44.	70.81
1.93	3.1	4.99	5.16	8.3	13.36	27.96	45.	72.42
1.99	3.2	5.15	5.22	8.4	13.52	28.58	46.	74.03
2.05	3.3	5.31	5.28	8.5	13.68	29.21	47.	75.64
2.11	3.4	5.47	5.34	8.6	13.84	29.83	48.	77.24
2.17	3.5	5.63	5.41	8.7	14.00	30.45	49.	78.85
2.24	3.6	5.79	5.47	8.8	14.16	31.07	50.	80.46
2.30	3.7	5.95	5.53	8.9	14.32	31.69	51.	82.07
2.36	3.8	6.12	5.59	9.0	14.48	32.31	52.	83.68
2.42	3.9	6.28	5.65	9.1	14.64	32.93	53.	85.29
2.49	4.0	6.44	5.72	9.2	14.81	33.56	54.	86.90
2.55	4.1	6.60	5.78	9.3	14.97	34.18	55.	88.51
2.61	4.2	6.76	5.84	9.4	15.13	34.80	56.	90.12
2.67	4.3	6.92	5.90	9.5	15.29	35.42	57.	91.73
2.73	4.4	7.08	5.97	9.6	15.45	36.04	58.	93.34
2.80	4.5	7.24	6.03	9.7	15.61	36.66	59.	94.95
2.86	4.6	7.40	6.09	9.8	15.77	37.28	60.	96.56
2.92	4.7	7.56	6.15	9.9	15.93	37.91	61.	98.17
2.98	4.8	7.72	6.21	10.0	16.09	38.53	62.	99.77
3.04	4.9	7.89	6.84	11.	17.70	39.15	63.	101.38
3.11	5.0	8.05	7.46	12.	19.31	39.77	64.	102.99
3.17	5.1	8.21	8.08	13.	20.92	40.39	65.	104.60
3.23	5.2	8.37	8.70	14.	22.53	41.01	66.	106.21

Table A-3

(continued)

MILES		KILOMETERS	MILES		KILOMETERS	MILES		KILOMETERS
41.63	67.	107.82	49.09	79.	127.13	56.55	91.	146.44
42.26	68.	109.43	49.71	80.	128.74	57.17	92.	148.05
42.88	69.	111.04	50.33	81.	130.35	57.79	93.	149.66
43.50	70.	112.65	50.95	82.	131.96	58.41	94.	151.27
44.12	71.	114.26	51.58	83.	133.57	59.03	95.	152.88
44.74	72.	115.87	52.20	84.	135.18	59.65	96.	154.49
45.36	73.	117.48	52.82	85.	136.79	60.28	97.	156.10
45.98	74.	119.09	53.44	86.	138.40	60.90	98.	157.71
46.61	75.	120.70	54.06	87.	140.01	61.52	99.	159.32
47.23	76.	122.30	54.68	88.	141.62	62.14	100.	160.93
47.85	77.	123.91	55.30	89.	143.22			
48.47	78.	125.52	55.93	90.	144.83			

Example: 6 miles = 9.66 kilometers
71 kilometers = 44.12 meters

Table A-4

CONVERSION --- MILES PER HOUR VS. METERS PER SECOND

MPH		m/sec	MPH		m/sec	MPH		m/sec
0.22	0.1	0.04	64.87	29.	12.96	147.64	66.	29.50
0.45	0.2	0.09	67.11	30.	13.41	149.87	67.	29.95
0.67	0.3	0.13	69.35	31.	13.86	152.11	68.	30.40
0.89	0.4	0.18	71.58	32.	14.31	154.35	69.	30.85
1.12	0.5	0.22	73.82	33.	14.75	156.59	70.	31.29
1.34	0.6	0.27	76.06	34.	15.20	158.82	71.	31.74
1.57	0.7	0.31	78.29	35.	15.65	161.06	72.	32.19
1.79	0.8	0.36	80.53	36.	16.09	163.30	73.	32.63
2.01	0.9	0.40	82.77	37.	16.54	165.53	74.	33.08
2.24	1.	0.45	85.00	38.	16.99	167.77	75.	33.53
4.47	2.	0.89	87.24	39.	17.43	170.01	76.	33.98
6.71	3.	1.34	89.48	40.	17.88	172.24	77.	34.42
8.95	4.	1.79	91.71	41.	18.33	174.48	78.	34.87
11.18	5.	2.24	93.95	42.	18.78	176.72	79.	35.32
13.42	6.	2.68	96.19	43.	19.22	178.95	80.	35.76
15.66	7.	3.13	98.43	44.	19.67	181.19	81.	36.21
17.90	8.	3.58	100.66	45.	20.12	183.43	82.	36.66
20.13	9.	4.02	102.90	46.	20.56	185.67	83.	37.10
22.37	10.	4.47	105.14	47.	21.01	187.90	84.	37.55
24.61	11.	4.92	107.37	48.	21.46	190.14	85.	38.00
26.84	12.	5.36	109.61	49.	21.90	192.38	86.	38.45
29.08	13.	5.81	111.85	50.	22.35	194.61	87.	38.89
31.32	14.	6.26	114.08	51.	22.80	196.85	88.	39.34
33.55	15.	6.71	116.32	52.	23.25	199.09	89.	39.79
35.79	16.	7.15	118.56	53.	23.69	201.32	90.	40.23
38.03	17.	7.60	120.79	54.	24.14	203.56	91.	40.68
40.26	18.	8.05	123.03	55.	24.59	205.80	92.	41.13
42.50	19.	8.49	125.27	56.	25.03	208.04	93.	41.57
44.74	20.	8.94	127.51	57.	25.48	210.27	94.	42.02
46.98	21.	9.39	129.74	58.	25.93	212.51	95.	42.47
49.21	22.	9.83	131.98	59.	26.38	214.75	96.	42.92
51.45	23.	10.28	134.22	60.	26.82	216.98	97.	43.36
53.69	24.	10.73	136.45	61.	27.27	219.22	98.	43.81
55.92	25.	11.18	138.69	62.	27.72	221.46	99.	44.26
58.16	26.	11.62	140.93	63.	28.16	223.69	100.	44.70
60.40	27.	12.07	143.16	64.	28.61			
62.63	28.	12.52	145.40	65.	29.06			

Example: 60 mph = 26.82 m/sec
5 m/sec = 11.18 mph

Table A-5

CONVERSION --- MILLIBARS VS. INCHES WATER COLUMN

MILLIBARS		INCHES W.C.	MILLIBARS		INCHES W.C.	MILLIBARS		INCHES W.C.
0.01	0.01	0.02	0.23	0.44	0.84	0.46	0.87	1.66
0.01	0.02	0.04	0.24	0.45	0.86	0.46	0.88	1.68
0.02	0.03	0.06	0.24	0.46	0.88	0.47	0.89	1.70
0.02	0.04	0.08	0.25	0.47	0.90	0.47	0.90	1.72
0.03	0.05	0.10	0.25	0.48	0.92	0.48	0.91	1.74
0.03	0.06	0.11	0.26	0.49	0.94	0.48	0.92	1.76
0.04	0.07	0.13	0.26	0.50	0.95	0.49	0.93	1.78
0.04	0.08	0.15	0.27	0.51	0.97	0.49	0.94	1.80
0.05	0.09	0.17	0.27	0.52	0.99	0.50	0.95	1.81
0.05	0.10	0.19	0.28	0.53	1.01	0.50	0.96	1.83
0.06	0.11	0.21	0.28	0.54	1.03	0.51	0.97	1.85
0.06	0.12	0.23	0.29	0.55	1.05	0.51	0.98	1.87
0.07	0.13	0.25	0.29	0.56	1.07	0.52	0.99	1.89
0.07	0.14	0.27	0.30	0.57	1.09	0.52	1.00	1.91
0.08	0.15	0.29	0.30	0.58	1.11	1.05	2.	3.82
0.08	0.16	0.31	0.31	0.59	1.13	1.57	3.	5.73
0.09	0.17	0.32	0.31	0.60	1.15	2.09	4.	7.64
0.09	0.18	0.34	0.32	0.61	1.17	2.62	5.	9.55
0.10	0.19	0.36	0.32	0.62	1.18	3.14	6.	11.46
0.10	0.20	0.38	0.33	0.63	1.20	3.67	7.	13.37
0.11	0.21	0.40	0.34	0.64	1.22	4.19	8.	15.28
0.12	0.22	0.42	0.34	0.65	1.24	4.71	9.	17.19
0.12	0.23	0.44	0.35	0.66	1.26	5.24	10.	19.10
0.13	0.24	0.46	0.35	0.67	1.28	10.47	20.	38.20
0.13	0.25	0.48	0.36	0.68	1.30	15.71	30.	57.30
0.14	0.26	0.50	0.36	0.69	1.32	20.94	40.	76.39
0.14	0.27	0.52	0.37	0.70	1.34	26.18	50.	95.49
0.15	0.28	0.53	0.37	0.71	1.36	31.42	60.	114.59
0.15	0.29	0.55	0.38	0.72	1.38	36.65	70.	133.69
0.16	0.30	0.57	0.38	0.73	1.39	41.89	80.	152.79
0.16	0.31	0.59	0.39	0.74	1.41	47.12	90.	171.89
0.17	0.32	0.61	0.39	0.75	1.43	52.36	100.	190.99
0.17	0.33	0.63	0.40	0.76	1.45	104.72	200.	381.97
0.18	0.34	0.65	0.40	0.77	1.47	157.08	300.	572.96
0.18	0.35	0.67	0.41	0.78	1.49	209.44	400.	763.94
0.19	0.36	0.69	0.41	0.79	1.51	261.80	500.	954.93
0.19	0.37	0.71	0.42	0.80	1.53	314.16	600.	1145.91
0.20	0.38	0.73	0.42	0.81	1.55	366.52	700.	1336.90
0.20	0.39	0.74	0.43	0.82	1.57	418.88	800.	1527.88
0.21	0.40	0.76	0.43	0.83	1.59	471.24	900.	1718.87
0.21	0.41	0.78	0.44	0.84	1.60	523.60	1000.	1909.85
0.22	0.42	0.80	0.45	0.85	1.62			
0.23	0.43	0.82	0.45	0.86	1.64			

Example: 4 millibars = 7.64 inches W.C.
0.33 inches W.C. = 0.17 millibars

Table A-6

CONVERSION --- MILLIBARS VS. INCHES OF MERCURY

MILLIBARS		INCHES Hg	MILLIBARS		INCHES Hg	MILLIBARS		INCHES Hg
3.	0.1	0.00	945.	32.	1.08	2126.	72.	2.44
6.	0.2	0.01	974.	33.	1.12	2156.	73.	2.47
9.	0.3	0.01	1004.	34.	1.15	2185.	74.	2.51
12.	0.4	0.01	1034.	35.	1.19	2215.	75.	2.54
15.	0.5	0.02	1063.	36.	1.22	2244.	76.	2.57
18.	0.6	0.02	1093.	37.	1.25	2274.	77.	2.61
21.	0.7	0.02	1122.	38.	1.29	2303.	78.	2.64
24.	0.8	0.03	1152.	39.	1.32	2333.	79.	2.68
27.	0.9	0.03	1181.	40.	1.35	2362.	80.	2.71
30.	1.0	0.03	1211.	41.	1.39	2392.	81.	2.74
59.	2.	0.07	1240.	42.	1.42	2421.	82.	2.78
89.	3.	0.10	1270.	43.	1.46	2451.	83.	2.81
118.	4.	0.14	1299.	44.	1.49	2481.	84.	2.84
148.	5.	0.17	1329.	45.	1.52	2510.	85.	2.88
177.	6.	0.20	1358.	46.	1.56	2540.	86.	2.91
207.	7.	0.24	1388.	47.	1.59	2569.	87.	2.95
236.	8.	0.27	1417.	48.	1.63	2599.	88.	2.98
266.	9.	0.30	1447.	49.	1.66	2628.	89.	3.01
295.	10.	0.34	1477.	50.	1.69	2658.	90.	3.05
325.	11.	0.37	1506.	51.	1.73	2687.	91.	3.08
354.	12.	0.41	1536.	52.	1.76	2717.	92.	3.12
384.	13.	0.44	1565.	53.	1.79	2746.	93.	3.15
413.	14.	0.47	1595.	54.	1.83	2776.	94.	3.18
443.	15.	0.51	1624.	55.	1.86	2805.	95.	3.22
472.	16.	0.54	1654.	56.	1.90	2835.	96.	3.25
502.	17.	0.58	1683.	57.	1.93	2864.	97.	3.28
532.	18.	0.61	1713.	58.	1.96	2894.	98.	3.32
561.	19.	0.64	1742.	59.	2.00	2923.	99.	3.35
591.	20.	0.68	1772.	60.	2.03	2953.	100.	3.39
620.	21.	0.71	1801.	61.	2.07	5906.	200.	6.77
650.	22.	0.75	1831.	62.	2.10	8859.	300.	10.16
679.	23.	0.78	1860.	63.	2.13	11812.	400.	13.55
709.	24.	0.81	1890.	64.	2.17	14765.	500.	16.93
738.	25.	0.85	1919.	65.	2.20	17718.	600.	20.32
768.	26.	0.88	1949.	66.	2.24	20671.	700.	23.70
797.	27.	0.91	1979.	67.	2.27	23624.	800.	27.09
827.	28.	0.95	2008.	68.	2.30	26577.	900.	30.48
856.	29.	0.98	2038.	69.	2.34	29530.	1000.	33.86
886.	30.	1.02	2067.	70.	2.37			
915.	31.	1.05	2097.	71.	2.40			

Example: 12 millibars = 0.41 inches Hg
98 inches Hg = 2894. millibars

Table A-7

CONVERSION --- MILLIBARS VS. POUNDS PER SQUARE INCH

MILLIBARS		PSI	MILLIBARS		PSI	MILLIBARS		PSI
1.	0.1	0.01	450.	31.	2.14	1030.	71.	4.90
3.	0.2	0.01	464.	32.	2.21	1044.	72.	4.97
4.	0.3	0.02	479.	33.	2.28	1059.	73.	5.03
6.	0.4	0.03	493.	34.	2.34	1073.	74.	5.10
7.	0.5	0.03	508.	35.	2.41	1088.	75.	5.17
9.	0.6	0.04	522.	36.	2.48	1102.	76.	5.24
10.	0.7	0.05	537.	37.	2.55	1117.	77.	5.31
12.	0.8	0.06	551.	38.	2.62	1131.	78.	5.38
13.	0.9	0.06	566.	39.	2.69	1146.	79.	5.45
15.	1.0	0.07	580.	40.	2.76	1160.	80.	5.52
29.	2.	0.14	595.	41.	2.83	1175.	81.	5.59
44.	3.	0.21	609.	42.	2.90	1189.	82.	5.66
58.	4.	0.28	624.	43.	2.97	1204.	83.	5.72
73.	5.	0.34	638.	44.	3.03	1218.	84.	5.79
87.	6.	0.41	653.	45.	3.10	1233.	85.	5.86
102.	7.	0.48	667.	46.	3.17	1247.	86.	5.93
116.	8.	0.55	682.	47.	3.24	1262.	87.	6.00
131.	9.	0.62	696.	48.	3.31	1276.	88.	6.07
145.	10.	0.69	711.	49.	3.38	1291.	89.	6.14
160.	11.	0.76	725.	50.	3.45	1305.	90.	6.21
174.	12.	0.83	740.	51.	3.52	1320.	91.	6.28
189.	13.	0.90	754.	52.	3.59	1334.	92.	6.34
203.	14.	0.97	769.	53.	3.66	1349.	93.	6.41
213.	14.7	1.01	783.	54.	3.72	1363.	94.	6.48
218.	15.	1.03	798.	55.	3.79	1378.	95.	6.55
232.	16.	1.10	812.	56.	3.86	1392.	96.	6.62
247.	17.	1.17	827.	57.	3.93	1407.	97.	6.69
261.	18.	1.24	841.	58.	4.00	1421.	98.	6.76
276.	19.	1.31	856.	59.	4.07	1436.	99.	6.83
290.	20.	1.38	870.	60.	4.14	1450.	100.	6.90
305.	21.	1.45	885.	61.	4.21	2900.	200.	13.79
319.	22.	1.52	899.	62.	4.28	4350.	300.	20.69
334.	23.	1.59	914.	63.	4.34	5800.	400.	27.59
348.	24.	1.66	928.	64.	4.41	7250.	500.	34.48
363.	25.	1.72	943.	65.	4.48	8700.	600.	41.38
377.	26.	1.79	957.	66.	4.55	10150.	700.	48.28
392.	27.	1.86	972.	67.	4.62	11600.	800.	55.17
406.	28.	1.93	986.	68.	4.69	13050.	900.	62.07
421.	29.	2.00	1001.	69.	4.76	14500.	1000.	68.97
435.	30.	2.07	1015.	70.	4.83			

EXAMPLE: 21 mb = 1.45 psi
2 psi = 29. mb

Table A-8

CONVERSION --- POUND PER HOUR VS. GRAM PER SECOND

lb/hr		gm/sec	lb/hr		gm/sec	lb/hr		gm/sec
0.08	0.01	0.001	182.38	23.	2.90	499.56	63.	7.95
0.16	0.02	0.003	190.31	24.	3.03	507.49	64.	8.07
0.24	0.03	0.004	198.24	25.	3.15	515.42	65.	8.20
0.32	0.04	0.005	206.17	26.	3.28	523.35	66.	8.32
0.40	0.05	0.006	214.10	27.	3.41	531.28	67.	8.45
0.48	0.06	0.008	222.03	28.	3.53	539.21	68.	8.58
0.56	0.07	0.009	229.96	29.	3.66	547.14	69.	8.70
0.63	0.08	0.010	237.89	30.	3.78	555.07	70.	8.83
0.71	0.09	0.011	245.81	31.	3.91	563.00	71.	8.95
0.79	0.1	0.013	253.74	32.	4.04	570.93	72.	9.08
1.59	0.2	0.025	261.67	33.	4.16	578.85	73.	9.21
2.38	0.3	0.038	269.60	34.	4.29	586.78	74.	9.33
3.17	0.4	0.050	277.53	35.	4.41	594.71	75.	9.46
3.96	0.5	0.063	285.46	36.	4.54	602.64	76.	9.58
4.76	0.6	0.076	293.39	37.	4.67	610.57	77.	9.71
5.55	0.7	0.088	301.32	38.	4.79	618.50	78.	9.84
6.34	0.8	0.101	309.25	39.	4.92	626.43	79.	9.96
7.14	0.9	0.114	317.18	40.	5.04	634.36	80.	10.09
7.93	1.	0.126	325.11	41.	5.17	642.29	81.	10.22
17.44	2.	0.25	333.04	42.	5.30	650.22	82.	10.34
23.79	3.	0.38	340.97	43.	5.42	658.15	83.	10.47
31.72	4.	0.50	348.90	44.	5.55	666.08	84.	10.59
39.65	5.	0.63	356.83	45.	5.68	674.01	85.	10.72
47.58	6.	0.76	364.76	46.	5.80	681.94	86.	10.85
55.51	7.	0.88	372.69	47.	5.93	689.87	87.	10.97
63.44	8.	1.01	380.62	48.	6.05	697.80	88.	11.10
71.37	9.	1.14	388.55	49.	6.18	705.73	89.	11.22
79.30	10.	1.26	396.48	50.	6.31	713.66	90.	11.35
87.22	11.	1.39	404.41	51.	6.43	721.59	91.	11.48
95.15	12.	1.51	412.33	52.	6.56	729.52	92.	11.60
103.08	13.	1.64	420.26	53.	6.68	737.44	93.	11.73
111.01	14.	1.77	428.19	54.	6.81	745.37	94.	11.85
118.94	15.	1.89	436.12	55.	6.94	753.30	95.	11.98
126.87	16.	2.02	444.05	56.	7.06	761.23	96.	12.11
134.80	17.	2.14	451.98	57.	7.19	769.16	97.	12.23
142.73	18.	2.27	459.91	58.	7.31	777.09	98.	12.36
150.66	19.	2.40	467.84	59.	7.44	785.02	99.	12.49
159.59	20.	2.52	475.77	60.	7.57	792.95	100.	12.61
166.52	21.	2.65	483.70	61.	7.69			
174.45	22.	2.77	491.63	62.	7.82			

Example: 47 lb/hr = 5.93 gm/sec
86 gm/sec = 681.94 lb/hr

Table A-9

CONVERSION --- DEGREES KELVIN TO DEGREES FARENHEIT

FARENHEIT	KELVIN	FARENHEIT	KELVIN	FARENHEIT	KELVIN	FARENHEIT	KELVIN
-40	233	11	261	62	290	113	318
-39	234	12	262	63	290	114	319
-38	234	13	262	64	291	115	319
-37	235	14	263	65	291	116	320
-36	235	15	264	66	292	117	320
-35	236	16	264	67	292	118	321
-34	236	17	265	68	293	119	321
-33	237	18	265	69	294	120	322
-32	237	19	266	70	294	121	322
-31	238	20	266	71	295	122	323
-30	239	21	267	72	295	123	324
-29	239	22	267	73	296	124	324
-28	240	23	268	74	296	125	325
-27	240	24	269	75	297	126	325
-26	241	25	269	76	297	127	326
-25	241	26	270	77	298	128	326
-24	242	27	270	78	299	129	327
-23	242	28	271	79	299	130	327
-22	243	29	271	80	300	131	328
-21	244	30	272	81	300	132	329
-20	244	31	272	82	301	133	329
-19	245	32	273	83	301	134	330
-18	245	33	274	84	302	135	330
-17	246	34	274	85	302	136	331
-16	246	35	275	86	303	137	331
-15	247	36	275	87	304	138	332
-14	247	37	276	88	304	139	332
-13	248	38	276	89	305	140	333
-12	249	39	277	90	305	141	334
-11	249	40	277	91	306	142	334
-10	250	41	278	92	306	143	335
- 9	250	42	279	93	307	144	335
- 8	251	43	279	94	308	145	336
- 7	251	44	280	95	308	146	336
- 6	252	45	280	96	309	147	337
- 5	252	46	281	97	309	148	337
- 4	253	47	281	98	310	149	338
- 3	254	48	282	99	310	150	339
- 2	254	49	282	100	311	151	339
- 1	255	50	283	101	311	152	340
0	255	51	284	102	312	153	340
1	256	52	284	103	312	154	341
2	256	53	285	104	313	155	341
3	257	54	285	105	314	156	342
4	257	55	286	106	314	157	342
5	258	56	286	107	315	158	343
6	259	57	287	108	315	159	344
7	259	58	287	109	316	160	344
8	260	59	288	110	316	161	345
9	260	60	289	111	317	162	345
10	261	61	289	112	317	163	346

Table A-9

(continued)

FARENHEIT	KELVIN	FARENHEIT	KELVIN	FARENHEIT	KELVIN	FARENHEIT	KELVIN
164	346	260	400	500	533	780	689
165	347	265	402	505	536	790	694
166	347	270	405	510	539	800	700
167	348	275	408	515	541	810	705
168	349	280	411	520	544	820	711
169	349	285	414	525	547	830	716
170	350	290	416	530	550	840	722
171	350	295	419	535	552	850	727
172	351	300	422	540	555	860	733
173	351	305	425	545	558	870	739
174	352	310	427	550	561	880	744
175	352	315	430	555	564	890	750
176	353	320	433	560	566	900	755
177	354	325	436	565	569	910	761
178	354	330	439	570	572	920	766
179	355	335	441	575	575	930	772
180	355	340	444	580	577	940	777
181	356	345	447	585	580	950	783
182	356	350	450	590	583	960	789
183	357	355	452	595	586	970	794
184	357	360	455	600	589	980	800
185	358	365	458	605	591	990	805
186	359	370	461	610	594	1000	811
187	359	375	464	615	597	1050	839
188	360	380	466	620	600	1100	866
189	361	385	469	625	602	1150	894
190	361	390	472	630	605	1200	922
191	361	395	475	635	608	1250	950
192	362	400	477	640	611	1300	977
193	363	405	480	645	614	1350	1005
194	363	410	483	650	616	1400	1033
195	364	415	486	655	619	1450	1061
196	364	420	489	660	622	1500	1089
197	365	425	491	665	625	1550	1116
198	365	430	494	670	627	1600	1144
199	366	435	497	675	630	1650	1172
200	366	440	500	680	633	1700	1200
205	369	445	502	685	636	1750	1227
210	372	450	505	690	639	1800	1255
215	375	455	508	695	641	1850	1283
220	377	460	511	700	644	1900	1311
225	380	465	514	710	650	1950	1339
230	383	470	516	720	655	2000	1366
235	386	475	519	730	661		
240	389	480	522	740	666		
245	391	485	525	750	672		
250	394	490	527	760	677		
255	397	495	530	770	683		

Example: 970 F = 794 K
338 K = 149 F

Glossary

Abatement. The reduction in degree or intensity of pollution.

Abrasion. The wearing away of surface material by the scouring action of moving solids, liquids, or gases.

Absorption. The penetration of one substance into or through another.

Acidity. The quantitative capacity of aqueous solutions to react with hydroxyl ions. It is measured by titration with a standard solution of a base to a specified end point. Usually expressed as milligrams per liter of calcium carbonate.

Activated Carbon. A highly adsorbent form of carbon used to remove odors and toxic substances from gaseous emissions or to remove dissolved organic material from wastewater.

Adequately Wetted. Sufficiently mixed or coated with water or an aqueous solution to prevent dust emissions.

Adhesion. Molecular attraction which holds the surfaces of two substances in contact, such as water and rock particles.

Adsorption. The attachment of the molecules of a liquid or gaseous substance to the surface of a solid.

Advanced Waste Water Treatment. The tertiary stage of sewage treatment.

Aeration. The act of exposing a liquid to air (oxygen) with the aim of producing a high level of dissolved oxygen in the liquid.

Aerosol. A particle of solid or liquid matter that can remain suspended in the air because of its small size.

Afterburner. A device that includes an auxiliary fuel burner and combustion chamber to incinerate combustible air contaminants.

Air Emissions. For stationary sources, the release or discharge of a pollutant by an owner or operator into the ambient air either by means of a stack or as a fugitive dust, mist, or vapor as a result inherent to the manufacturing or forming process.

Air Pollutant. Dust, fumes, smoke, and other particulate matter, vapor, gas, odorous substances, or any combination thereof. Also any air pollution agent or combination of such agents, including any physical, chemical, biological, radioactive substance or matter which is emitted into or otherwise enters the ambient air.

Air Pollution. The presence of any air pollutant in sufficient quantities and of such characteristics and duration as to be, or likely to be, injurious to health or welfare, animal or plant life, or property, or as to interfere with the enjoyment of life or property.

Alkalinity. The measurable ability of solutions or aqueous suspensions to neutralize an acid.

Ambient Air. That portion of the atmosphere external to buildings to which the general public has access.

Ash. Inorganic residue remaining after ignition of combustible substances determined by definite prescribed methods.

Autogenous (Autothermic) Combustion. The burning of wet organic material where the moisture content is at such a level that the heat of combustion of the organic material is sufficient to vaporize the water and maintain combustion where no auxiliary fuel is required except for start-up of the process.

Background Level. With respect to air pollution, amounts of pollutants present in the ambient air due to natural sources.

Baffles. Deflector vanes, guides, grids, grating, or similar devices constructed or placed in air or gas flow systems, flowing water, or slurry systems to effect a more uniform distribution of velocities, absorb energy, divert, guide, or agitate fluids and check eddies.

Baghouse. An air pollution abatement device used to trap particulates by filtering gas streams through large fabric bags usually made of cloth or glass fibers.

Barometric Damper. A hinged or pivoted plate that automatically regulates the amount of air entering a duct, breeching, flue connection, or stack, thereby maintaining a constant draft within an incinerator.

Barrel. A measure of 42 gallons at 60°F.

Batch-Fed Incinerator. An incinerator that is periodically charged with waste, and where one charge is allowed to burn out before another is added.

Biomass. The amount of living matter in a given unit of the environment.

Biosphere. The portion of the Earth and its atmosphere that can support life.

Biota. All living organisms that exist in an area.

Blower. A fan used to force air or gas under pressure.

Bottom Ash. The solid material that remains on a hearth or falls off the grate after thermal processing is complete.

Breeching. A passage that conducts the products of combustion to a stack or chimney.

Bridge Wall. A partition between chambers over which the products of combustion pass.

British Thermal Unit (Btu). The amount of heat required to raise the temperature of 1 pound of water 1°F.

Burning Hearth. A solid surface to support the solid fuel or solid waste in a furnace and upon which materials are placed for combustion.

Burning Rate. The volume of solid waste incinerated or the amount of heat released during incineration. The burning rate is usually expressed in pounds of solid waste per square foot of burning area per area or in Btus per cubic foot of furnace volume per hour.

Butterfly Damper. A plate or blade installed in a duct, breeching, flue connection, or stack that rotates on an axis to regulate the flow of gases.

Carbonaceous Matter. Pure carbon or carbon compounds present in the fuel or residue of a combustion process.

Carbon Sorption. The process in which a substance (the sorbate) is brought into contact with a solid (the sorbent), usually activated carbon, and held there either by chemical or physical means.

Carcinogenic. Capable of causing the cells of an organism to react in such a way as to produce cancer.

Catalytic Combustion System. A process in which a substance is introduced into an exhaust gas stream to burn or oxidize vaporized hydrocarbons or odorous contaminants, while the substance itself remains intact.

Caustic Soda. Sodium Hydroxide ($NaOH$), a strong alkaline substance used as the cleaning agent in some detergents.

Chemical Oxygen Demand. A measure of the oxygen-consuming capacity of inorganic and organic matter present in water or wastewater. It is expressed as the amount of oxygen consumed from a chemical oxidant in a specific test. It does not differentiate between stable and unstable organic matter and thus does not necessarily correlate with biochemical oxygen demand.

Classification. The separation and rearrangement of waste materials according to composition (e.g., organic or inorganic), size, weight, color, shape, and the like, using specialized equipment.

Clinkers. Hard, sintered, or fused pieces of residue formed in a furnace by the agglomeration of ash, metals, glass, and ceramics.

Combustibles. Materials that can be ignited at a specific temperature in the presence of air to release heat energy.

Combustion. The production of heat and light energy through a chemical process, usually oxidation.

Combustion Air. The air used for burning a fuel.

Combustion Gases. The mixture of gases and vapors produced by burning.

Contaminant. Any physical, chemical, biological, or radiological substance in water.

Continuous Feed Incinerator. An incinerator into which solid waste is charged almost continuously to maintain a steady rate of burning.

Controlled-Air Incinerator. An incinerator with two or more combustion chambers in which the amounts and distribution of air are controlled. Partial combustion takes place in the first zone and gases are burned in a subsequent zone or zones.

Corrosion. The gradual wearing away of a substance by chemical action.

Curtain Wall. A refractory construction or baffle that deflects combustion gases downward.

Cyclone Separator. A separator that uses a swirling air flow to sort mixed materials according to the size, weight and density of the pieces.

Deliquescent. The ability to absorb water from the air.

Discharge. The release of any waste stream or any constituent thereof to the environment.

Dispersion Technique. The use of dilution to attain ambient air quality levels including any intermittent or supplemental control of air pollutants varying with atmospheric conditions.

Disposal. The planned release or placement of waste in a manner that precludes recovery.

Draft. The difference between the pressure within an incinerator and that in the atmosphere.

Drying Hearth. A solid surface in an incinerator upon which wet waste materials, liquids, or waste matter that may turn to liquid before burning are placed to dry or to burn with the help of hot combustion gases.

Duct. A conduit, usually metal or fiberglass, round or rectangular in cross section, used for conveyance of air.

Dust. Fine grain particles light enough to be suspended in the air.

Effluent. Waste materials, usually waterborne, discharged into the environment, treated or untreated.

Electrostatic Precipitator (ESP). An air pollution control device that imparts an electric charge to particles in a gas stream causing them to collect on an electrode.

Emission Rate. The amount of pollutant emitted per unit of time.

Environment. Water, air, land, all plants, human beings, and other animals living therein, and the interrelationships which exist among them.

Environmental Impact Statement. A document required of Federal agencies by the National Environmental Policy Act for major projects or legislative proposals.

Exhaust System. The system comprised of a combination of components which provides for enclosed flow of exhaust gas from the furnace exhaust port to the atmosphere.

Extraction Test Procedure. A series of laboratory operations and analyses designed to determine whether, under severe conditions, a solid waste, stabilized waste, or landfill material can yield a hazardous leachate.

Federal Register. Daily publication that is the official method of notice of executive branch actions, including proposed and final regulations.

Filter. A porous device through which a gas or liquid is passed to remove suspended particles or dust.

Firebrick. Refractory brick made from fireclay.

Fixed Carbon. The ash-free carbonaceous material that remains after volatile matter is driven off a dry solid waste sample.

Fixed Grate. A grate without moving parts, also called a stationary grate.

Flammable Waste. A waste capable of igniting easily and burning rapidly.

Flash Drying. The process of drying a wet organic material by passing it through a high temperature zone at such a rate that the water is rapidly evaporated but the organic material, protected by the boiling point of water, is not overheated.

Flue. Any passage designed to carry combustion gases and entrained particulates.

Flue-Fed Incinerator. An incinerator that is charged through a shaft that functions as a chute for charging waste ash as a flue for carrying the products of combustion.

Flue Gas. The products of combustion, including pollutants, emitted to the air after a production process or combustion takes place.

Fluidized Bed Combustion. Oxidation of combustible material within a bed of solid, inert (noncombustible) particles which under the action of vertical hot air flow will act as a fluid.

Fly Ash. The airborne combustion residue from burning fuel.

Forced Draft. The positive pressure created by the action of a fan or blower which supplies the primary or secondary combustion air in an incinerator.

Fugitive Emissions. Emissions other than those from stacks or vents.

Fume. Solid particles under 1μ in diameter, formed as vapors condense or as chemical reactions take place.

Furnace. A combustion chamber; an enclosed structure in which heat is produced.

Garbage. Solid waste resulting from animal, grain, fruit, or vegetable matter used or intended for use as food.

Grain Loading. The rate at which particles are emitted from a pollution source, in grains per cubic foot of gas emitted, where 7000 grains weigh 1 pound.

Grate. A piece of furnace equipment used to support solid waste or solid fuel during the drying, igniting, and burning process.

Guillotine Damper. An adjustable plate, used to regulate the flow of gases, installed vertically in a breeching.

Hazardous Waste. A waste, or combination of wastes, that may cause or significantly contribute to an increase in mortality or an increase in serious irreversible, or incapacitating reversible illness or that pose a substantial present or potential hazard to human health or the environment when improperly treated, stored, transported, disposed of, or otherwise managed.

Hearth. The bottom of a furnace on which waste materials are exposed to the flame.

Heat Balance. An accounting of the distribution of the heat input and output of an incinerator or boiler, usually on an hourly basis.

Heavy Metals. Metallic elements such as mercury, chromium, lead, cadmium, and arsenic, with high atomic weights, and which tend to accumulate in the food chain.

Hydrocarbon. Any of a vast family of compounds containing carbon and hydrogen in various combinations, found especially in fossil fuels.

Ignitability. Liquids having a flash point of less than 140°F, and non-liquids liable to cause fires through friction, absorption of moisture, spontaneous chemical change, or retained heat from manufacturing, or are liable, when ignited, to burn so vigorously and persistently as to create a hazard; ignitable compressed gas oxidizers.

Incineration. An engineered process using controlled flame combustion to thermally degrade waste materials.

Incinerator Stoker. A mechanically operable moving grate arrangement for supporting, burning, or transporting solid waste in a furnace and discharging the residue.

Induced Draft. The negative pressure created by the action of a fan, blower, or other gas-moving device located between an incinerator and a stack.

Infiltration Air. Air that leaks into the chambers or ducts of an incinerator.

Inorganic Matter. Chemical substances of mineral origin, not containing carbon-to-carbon bonding.

Isokinetic Sampling. Sampling in which the linear velocity of the gas entering the sampling nozzle is equal to that of the undisturbed gas stream at the sample point.

Landfill. A land disposal site employing an engineered method of disposing of wastes on land that minimizes environmental hazards by spreading wastes in thin layers, compacting the wastes to the smallest practical volume, and applying cover materials at the end of each operating day.

Lime. Any of a family of chemicals consisting essentially of calcium oxide or hydroxide made from limestone (calcite).

Masking. Blocking out one sight, sound, or smell with another.

Mist. Liquid particles, measuring 40 to 500μ in diameter, that are formed by condensation of vapor.

Mixing Chamber. A chamber usually placed between the primary and secondary combustion chambers and in which the products of combustion are thoroughly mixed by turbulence that is created by increased velocities of gases, checker work, or turns in the direction of the gas flow.

Modular Combustion Unit. One of a series of incinerator units designed to operate independently and that can handle small quantities of solid waste.

Multiple Chamber Incinerator. An incinerator that consists of two or more chambers, arranged as in-line or retort types, interconnected by gas passage parts or flues.

Municipal Incinerator. A privately or publicly owned incinerator primarily designed and used to burn residential and commercial solid waste within a community.

Municipal Solid Wastes. Garbage, refuse, sludges, and other discarded materials resulting from residential and nonindustrial operations and activities.

Mutagenic. The property of a substance or mixture of substances which, when it interacts with a living organism, causes the genetic characteristics of the organism to change and its offspring to have a decreased life expectancy.

Natural Draft. The negative pressure created by the height of a stack or chimney and the difference in temperature between flue gases and the atmosphere.

Neutralization. The chemical process in which the acidic or basic characteristics of a fluid are changed to those of water.

Odor Threshold. The lowest concentration of an airborne odor that a human being can detect.

Opacity. Degree of obscuration of light; for example, a window has zero opacity while a wall has 100% opacity.

Open Pit Incinerator. A burning apparatus that has an open top and a system of closely spaced nozzles that place a stream of high velocity air over the burning zone.

Organic Matter. Chemical substances comprised mainly of carbon, covalently bonded. May have its origin in animal or plant life, coal, petroleum, or laboratory synthesis.

Organic Nitrogen. Nitrogen combined in organic molecules such as protein, amines, and amino acids.

Overfire Air. Air under control as to quantity and direction, introduced above and beyond a fuel bed by induced or forced draft.

Oxidant. A substance containing oxygen that reacts chemically to release some of or all of its oxygen component to another substance.

Packed Tower. A pollution control device that forces dirty gas through a tower packed with crushed rock, wood chips, or other packing while liquid is sprayed over the packing material. Pollutants in the gas stream either dissolve in or chemically react with the liquid.

Particulate Matter. Any material, except water in uncombined form, that is or has been airborne and exists as a liquid or a solid at standard conditions.

Particulates. Fine liquid or solid particles such as dust, smoke, mist, fumes, or smog, found in the air or in emissions.

Permeability. The property of a solid material which allows fluid to flow through it.

Plume. Visible emissions from a flue or chimney.

Pollutant. Dredged spoil, solid waste, incinerator residue, sewage, garbage, sewage sludge, munitions, chemical wastes, biological materials, radioactive materials, heat, wrecked or discarded equipment, rock, sand, cellar dirt, and industrial, municipal, and agricultural waste discharged into water.

Pollution. The presence of matter or energy whose nature, location, or quantity produces undesired environmental effects. Also, the man-made or man-induced alteration of the chemical, physical, biological, and radiological integrity of water.

Porosity. The ratio of the volume of pores of a material to the volume of its mass.

Precipitators. Air pollution control devices that collect particles from an emission by mechanical or electrical means.

Precursor. A pollutant that takes part in a chemical reaction resulting in the formation of one or more new pollutants.

Primary Combustion Air. The air admitted to a combustion system when the fuel is first oxidized.

Primary Pollutant. A pollutant emitted directly from a polluting stack.

Primary Standard. A national air emissions standard intended to establish a level of air quality that, with an adequate margin of error, will protect public health.

Putrescible. A substance that can rot quickly enough to cause odors and attract flies.

Pyrolysis. The chemical decomposition of organic matter through the application of heat in an oxygen deficient atmosphere.

Quench Tank. A water filled tank used to cool incinerator residues or hot materials during industrial processes.

Reactivity. The tendency to create vigorous reactions with air or water, tendency to explode, to exhibit thermal instability with regard to shock, ready reaction to generate toxic gases.

Refractory Material. Nonmetallic substances used to line furnaces because they can endure high temperatures and resist abrasion, spalling, and slagging.

Refuse. All solid materials which are discarded as useless.

Refuse Derived Fuel. The combustible, or organic, portion of municipal waste that has been separated out and processed for use as fuel.

Residue. Solid or semisolid materials such as, but not limited to, ash, ceramics, glass, metal, and organic substances remaining after incineration or processsing.

Scrubbing. The removal of impurities from a gas stream by spraying of a fluid.

Secondary Combustion Air. The air introduced above or below the fuel (waste) bed by a natural, induced, or forced draft.

Secondary Pollutant. A pollutant formed in the atmosphere by chemical changes taking place between primary pollutants and other substances present in the air.

Secondary Standard. A national air quality standard that establishes that ambient concentration of a pollutant that, with an adequate margin of safety, will protect the public welfare (all parts of the environment other than human health) from adverse impacts.

Settling Chamber. Any chamber designed to reduce the velocity of the products of combustion and thus to promote the settling of fly ash from the gas stream before it is discharged to the next process or to the environment.

Settling Velocity. The mean velocity at which a given dust particle will fall; also terminal velocity.

Slag. The more or less completely fused and vitrified matter separated during the reduction of metal from its ore.

Sludge. Any solid, semisolid, or liquid waste generated from a municipal, commercial, or industrial wastewater treatment plant, water supply treatment plant, or air pollution control facility, or any other such waste having similar characteristics and effects.

Smog. The irritating haze resulting from the sun's effect on certain pollutants in the air.

Smoke. Particles suspended in air after incomplete combustion of materials containing carbon.

Solid Waste. Any garbage, refuse, or sludge from a waste treatment plant, water supply treatment plant, or air pollution control facility and other discarded material, including solid, liquid, semisolid, or contained gaseous material resulting from industrial, commercial, mining, and agricultural operations, and from community activities.

Soot. Carbon dust formed by incomplete combustion.

Source. Any building, structure, facility, or installation from which there is or may be the discharge of pollutants.

Spray Chamber. A chamber equipped with water sprays that cool and clean the combustion products passing through it.

Stack. Any chimney, flue, vent, roof monitor, conduit, or duct arranged to discharge emissions to the ambient air.

Stationary Source. Any building, structure, facility, or installation which emits or may emit any air pollutant.

Stoichiometric Combustion. Combustion with the theoretical air quantity.

Stoker. A mechanical device to feed solid fuel or solid waste to a furnace.

Teratogenic. Affecting the genetic characteristics of an organism so as to cause its offspring to be misshapen or malformed.

Theoretical Air. The quantity of air, calculated from the chemical composition of a waste, that is required to burn the waste completely.

Toxic Substance. A chemical or mixture that may present an unreasonable risk of injury to health or to the environment.

Underfire Air. Forced or induced combustion air (quantity and direction are controlled) introduced under a grate to promote burning within a fuel bed.

Vapor. The gaseous phase of substances that are liquid or solid at atmospheric temperature and pressure—such as steam.

Vapor Plume. The stack effluent consisting of flue gases made visible by condensed water droplets or mist.

Volatile. Any substance that evaporates at a low temperature.

Volatility. The property of a substance or substances to convert into vapor or gas without chemical change.

Waste. Unwanted materials left over from manufacturing processes, refuse from places of human or animal habitation.

Waterwall Incinerator. An incinerator whose furnace walls consist of vertically arranged metal tubes through which water passes and absorbs the radiant energy from burning solid waste.

Index